Seeds: Their Conservation Principles & Practices

SEEDS: THEIR CONSERVATION PRINCIPLES & PRACTICES

Dr. Veena Gupta
Principal Scientist
National Bureau of Plant
Genetic Resources
New Delhi-110012

2009

SSPH **SATISH SERIAL PUBLISHING HOUSE**
403, Express Tower, Commercial Complex, Azadpur, Delhi - 110033 (INDIA)
Phone : 011-27672852 Fax : 91-11-27672046
E-mail : info@satishserial.com, hkjain1975@yahoo.com
Website : www.satishserial.com

Published by :

SATISH SERIAL PUBLISHING HOUSE
403, Express Tower, Commercial Complex, Azadpur, Delhi - 110033 (INDIA)
Phone : 011-27672852 Fax : 91-11-27672046
E-mail : info@satishserial.com, hkjain1975@yahoo.com

ISBN 81-89304-58-5

Printed at
SALASAR IMAGING SYSTEMS, DELHI-35

Preface

Global awareness of rapid loss of biodiversity is increasing manifolds. Biodiversity is of immense ecological, economic and cultural value for the existence of mankind. Developing countries in the tropical and subtropical regions of world are biodiversity rich while the developed part of world is technology rich and conserve the biodiversity including agro-biodiversity through various strategies viz. seed genebanks, in vitro banks, cryo banks and DNA Banks of all the methods of conservation, seed banks are most convenient and cheapest method common used, being both economical and practiced. The present book gives an holistic approaches in practical manner for the ex situ conservation of seed. Although immense interactive is available on several of the topics covered the present book gives a semi-technical approach to highlight the various issues dealing with biodiversity conservation specially the genebank procedures regarding physiological and biochemical procedures required for biodiversity management and conservation.

The book Seeds: Their Conservation, Principles & Practices is a practical technical guide for students and researchers engaged in biodiversity conservation programme. It has been interceded to provide emphasis on the detailed practical methods along with theoretical information and to serve as a ready reference source for students, research workers engaged in biodiversity conservation.

Recent development in view of Convention on Biological Diversity and World Trade Organistaion, and other related topics are also covering with special reference to Indian scenario in a very lucid style for students and researchers. All the illustration presented in the book are drawn by the author. The author acknowledge her dept to all those who shared their information and help in this endeavor.

New Delhi
May 2008

Veena Gupta
Principal Scientist

Contents

1

Biodiversity-Importance and Value

Biodiversity is defined by the Convention on Biological Diversity as 'the variability among living organisms from all sources, including terrestrial, marine and other aquatic ecosystems and the ecological complexes of which they are part, this includes diversity within species, between species and of ecosystems. More simply, biodiversity is the variety of the world's organisms, including their genetic makeup and the com munities they form. Biodiversity is dynamic: the genetic composition of species changes over time in response to natural and human-induced selection pressures; the occurrence and relative abundance of species in ecological community's changes as a result of ecological and physical factors. Biodiversity represents the very foundation of human existence. Yet by our heedless actions we are eroding this biological capital at an alarming rate. Even today, despite the destruction that we have inflicted on the environment and its natural bounty, its resilience is taken for granted. But the more we learn of the workings of the natural world, the clearer it becomes that there is a limit to the disruption that environment can endure. The entry into force of the Convention on Biological Diversity, in December 1993, is illustrative not only of this recognition but also of a change in the overall strategy in conserving biodiversity. It signals a move to a more proactive position that simultaneously seeks to meet people's needs from biological resources while ensuring the long-term sustainability of Earth's biological capital

The primary causes underlying the loss of biodiversity are demographic, economic, institutional, and technological factors, including:

- Increasing demands for biological resources due to increasing population and economic development;
- Failure of people to consider the long-term consequences of their actions, often due to a basic lack of knowledge;
- Failure of people to appreciate the consequences of using inappropriate technology;
- Failure of economic markets to recognize the true value of biodiversity;
- Failure of economic markets to apply the global values of biodiversity at local level;
- Institutional failure to regulate the use of biological resources resulting from the growth in urbanization, changes in property rights, and shifting cultural attitudes;
- Failure of government policies to address the over-use of biological resources; and
- Increasing human migration, travel, and international trade.

The specific mechanisms by which the underlying causal factors result in the reduction and loss of populations, the extinction of species and the transformation and degradation of ecological communities include:

- Habitat loss, fragmentation and degradation (or the transformation of ecological communities to other uses),
- Over-exploitation;
- The introduction of non native species;
- Pollution or toxification of the soil, water and atmosphere;
- Climate change.

A wide variety of measures can be used to conserve biodiversity, including both in situ and ex situ methods.

Ecological diversity : Ecological systems do not exist as discrete units, but repre sent different parts of a natural continuum. Although terms such as forest, grassland, wet land and coral reef are commonly used to denote ecological systems, their delineation and spatial scale often depend on the intended purpose of the classification.

Some commonly used terms are:

Biome: a continental-scale region characterized by its distinctive vegetation and climate

Ecosystem: the individuals, populations and species that occur in a defined area, including their interactions with each other and with their physical environment.

Ecological community: a group of species inhabiting a particular area. Habitat -the biological and physical environment of a particular species.

Organismal diversity: The total number of species on Earth is estimated at between 13 and 14 million, of which only 1.75 million have been described. The enor mous diversity between these species, ranging from common annual herbs to bacteria of deep ocea n trenches, their arrangement into classifications reflecting their phyletic relationships, and the complex patterns of variation and distribution that they show, provide the very substance of biodiversity. Groups of interbreeding individuals within a species form distinct populations. Groups such as plants, birds, mammals, fishes, reptiles and amphibians the species with which we are most familiar account for only 3% of the estimated total, while the majority of species belong to groups such as insects, arachnids, fungi, nematodes and microorganisms.

Genetic Diversity : Genetic differences between the individuals of a species provide the basis for the diversity that is found between species. Molecular studies have revealed a wealth of genetic variability in most species: in fact, individuals of vir tually all species are genetically unique. Genetic diversity can be described at multiple levels from single genes to visible multi-locus traits. It is expressed as genetic variability both within and between populations. The amount and distribution of genetic diversity vary extensively among species in ways which are incompletely understood. It is well established, however, that genetic diversity within species is necessary to allow them to adapt to changing environmental conditions.

Agro- biodiversity is an evolutionarily divergent and highly inter-related component of biodiversity dealing with agro-ecosystems, and variation in agriculture related plants, animals, fish, insects, avians and microbes, etc. Agro-biodiversity is a minute fraction of our total global biodiversity (Fig. 1). This component of biodiversity is the basis of world's food supply, farm livelihoods and landscapes; and is humanity's insurance against future threats to food and farming. Crop and livestock diversity within species is as important as diversity between species. The second group of plant genetic resources belongs to other species. The distinctive features of agricultural biodiversity as compared to other components of biodiversity are:

- Conservation of agricultural biodiversity is inherently linked to its sustainable use and is actively managed by farmers.
- There is a high degree of dependence on human intervention for its survival and inter-dependence between countries for genetic resources for food and agriculture.

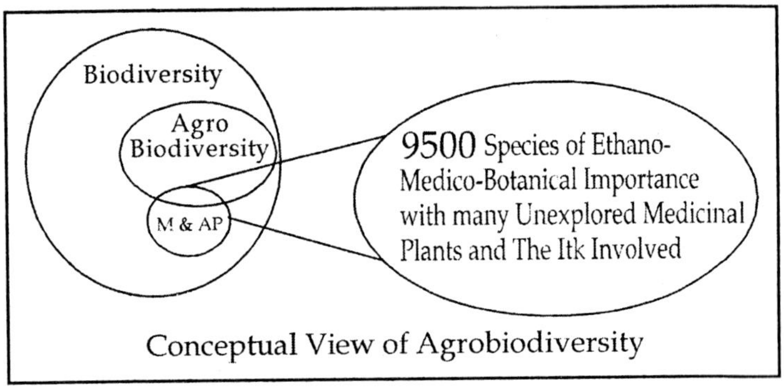

Fig.1. Biodiversity vs Agro-biodiversity

Agrobiodiversity vis-à-vis genetic variability is important for three following reasons:

- Providing stability to farming systems by incorporating a range of inter and intra-specific diversity. Under such systems, losses suffered by one variety can be minimized by continuity and gains of other.

- Providing insurance against anticipated changes in environmental conditions, biotic or abiotic stresses or climates.
- Embodying characteristics that are potentially valuable, but not yet exploited.

Centers of plant genetic diversity

There is a wide variation in the estimation of the total number of species or organisms existing: depending on different methodologies the estimates vary from 5 million to more than 360 million species. The Global Biodiversity Assessment published in 1995 by the United Nations Environment Programme estimates that about 13 to 14 million species might exist on our planet. Of this, less than 2 million have so far been scientifically described. An estimated 248,000 higher plant species are listed worldwide, of these fewer than 10% could be considered to be of economic value, and perhaps 500 would account for almost all economically valuable plants, only 150 species are commercially important. Globally, 90% of our food requirements come from 15-20 plants and animals although 10, 000 plant species have been cultivated over time. Some 40% of the world's market economy is based upon biological products and processes. In the rural communities of Asia, Latin America and Africa where majority of world population lives the dependence on biological products can be over 90% of human survival requirement. The value of the world's biological diversity has been now recognized by the international community. This has had a "cause and effect" kind of impact on the recent focus on economic valuation of plant genetic resources; as a result, genetic resources are now becoming an international issue.

The known plant species are not evenly distributed; some geographic regions are more important source of germplasm than others. According to OECD, (1996) species richness tends to be higher in warm and wet regions than in cold and dry regions and in areas with more variation in seasons and topography than in areas with less of variation in these. Regions dominated by developing countries which have more of tropical and subtropical areas are more species rich, for example, nearly 30% of all higher plant species are found in South America whereas, less than 5% in European and North American countries (Virchow, 1999).

In addition to the differential regional distribution of species, significant plant diversity can be located in specific regions where those species domesticated and evolved. The Russian botanist and geneticist

N.I. Vavilov defined the regions with the highest diversity within a crop as the center of origin and diversity of that crop plant (Vavilov, 1926). He identified eight major centers of origin and diversity: East Asia, tropical Asia, South Western Asia, the near East, Mediterranean, Abyssinnia, Andean and Central America. Different species of the same crop, however, have been domesticated in different regions and likewise same species of a crop was domesticated independently in different places. In addition, evolutionary forces along with human activities have led to formation of secondary centers of diversity (Harlan, 1971). Zeven and Zhukovsky (1975) enlarged Vavilov's 8 centers of origin and diversity by secondary center of diversity and drew the mega centers of cultivated plants to describe the origin and the main region of the present diversity of major crops on global level. The centers of origin of a crop plant may or may not be its important areas of production. Secondary centers of diversity have formed where significant diversity in crops (originated else where) has flourished.

The Indian gene center

The Indian subcontinent occupies a unique position among major gene rich countries of the world and a large number of crops are grown here. It is one of the 12 identified centers of origin of cultivated plants. Its wide agro-climatic, geophysical and ethnic diversity and its location at the confluence of three biogeography realms (Paleo-arctic, Afro-tropical and Indo-Malayan) contribute towards making India one of 12 mega-diversity zones of the world. It has also got two hot spots areas for biodiversity, one being in the Northeast and the other in Western Ghats. There are 16 forest types and 251 sub-types in India. The region has been divided into 21 agro-ecological zones, which offer nearly every ecological situation in the world, and which are separated by natural features and crop growing periods. Each of these agro-ecological zones is in turn comprised of myriad micro-habitats.

Over the millennia farming communities have selectively domesticated, exploited and nurtured the wild diversity gene pools under these diverse situations. The naturally occurring variation selected and maintained by farm-women and men was maintained in equilibrium with the diverse environmental conditions and met different human needs. The tribal-inhabited belt is more often the center of domestication and genetic diversity of food crops (cereals and pseudocereals, millets, grain legumes, vegetables, spices and condiments, oil seeds etc.), being maintained by peasants and subsistence farmers. These areas hold unique and important genetic materials which should be strictly protected against heavy grazing, intensive farming, commercial logging, construction of highways, dams and hydro-electric stations, resettlement projects, mining operations, etc. These are a few good examples of conscious folk domestication/ conservation (tribal domestication of a tuberous root legume soh-phlong *Moghania vestita* was brought under cultivation by the Khasi & Jaintia tribals, tubers are eaten raw. The tribal communities hold an array of potential plants diversity ranging from wild, semi-wild and domesticated forms and related information and indigenous knowledge about their uses. Thus, native plant genetic wealth, by the large, is preserved by different ethnic tribes who still live in isolation, are culturally distinct from the rural and urban communities and hold PGR in their backyards, kitchen gardens, marginal lands and near compounds in protected state. Due to diverse agro-ecological conditions and agro-climatic situations and agricultural practices, a number of crops species have developed rich diversity as cultivars/ landraces among the native as well as introduced types. The floristic diversity in wild relatives of cultivated types, wild valuable species and weedy taxa form an integral part of native wealth and constitute a useful gene pool. "Secondary crops" like limabean, velvetbean, winged bean, rice bean, have been shown to have reasonably high productivities without need for subsequent management inputs. Since minor species tend to be quickly lost with the adoption and expansion of more sophisticated farming systems there is an urgent need to systematically collect and maintain indigenous germplasm.

These ecosystems are subject to modifications that may substantially change their ability to support biodiversity. Even though there is an increased emphasis on identification of rare or under-exploited plants of potential value due to the emergence of sustainability concerns about the present day agriculture which has excessive reliance on mono-cropping and intensive use of inputs including agro-chemicals, with the overall objectives of having wider genetic base of agriculture. The loss of world's biological diversity and its economic and ecological consequences are now widely recognized as an environmental matter of urgent global concern. Several international developments in the recent past like the Convention on Biological Diversity have given momentum in realizing the importance of biological resources as base material for any country's economic growth and development.

Valuation of biodiversity

Biodiversity has always provided humanity with food, fuel, clothing, building materials and medicines, but recent advances in science and technology have greatly expanded the range of products and commercially valuable information that can be derived from genetic resources. Bioprospecting- the intensive search for novel secondary metabolites/products from natural sources, microorganisms, plants and animals for human use- has been carried out since prehistoric times. Individual genes, enzymes, and other biochemical compounds have become an important source of innovation in agrochemical, cosmetic and pharmaceutical industries. The monetary cost of each plant extinction could be viewed in terms of lost opportunities for development/isolation of new gene, enzyme, drug, biochemical compound.

Economic benefits have been derived from higher crop yields, fewer crop losses, diseases and pest resistance and greater ecological tolerance. Genes from different species could be transferred to improve crop yields; to improve resistance to disease, low humidity, high salinity, temperature extremes and pesticides; to increase concentration of nutraceuticals; to produce recombinant seeds, biological control agents and novel bioconversion routes. New and richer sources of food than existing food crops may be found in various exotic wild species. Most genetic improvements in economically important plants and animals in the past have come through genetic diversity within species rather than diversity between or across species although later have become increasingly more important with advent of genetic engineering. Over the last 100 years there has been a dramatic growth in agriculture productivity; world crop yield increased between 2-4 fold depending on the crop. An estimated 20-40% of this increased has been achieved by genetic modification and breeding. A fair proportion of yield gains have been related to the size and accessibility of germplasm collections, for example, Evenson and Gollin(1997) attributed 5.6% of India's yield gains in paddy to genetic resource collections. However, larger collections do not necessarily result in more economic improvements. Although gene banks are not the sole source of insurance against crop losses, they reduce expected future crop losses. This benefit is weighed against the cost of operating a genebank and of searching across the collection of germplasm.

It has been estimated that scientist effectively employ only 2% of the approximately 6 million accessions of plant species conserved in germplasm banks worldwide in the development of new varieties. The maintenance of plant genetic resources is a socio-economic activity. Since we cannot afford to preserve and collect most / all species and it is difficult, if not impossible, to decide whether all species are of equal or similar value, prioritization of genetic resources based on some criteria would help to direct collection and conservation activities. The estimation of plant genetic resources values is relatively new for economists. There is no direct market for plant genetic resources except for crop varieties. The market price of germplasm is not an appropriate indicator of its value because it does not normally reflect in full all of actual or potential benefits derivable from genetic material specifically, benefit from current use, future use options, and benefits from existence per se (Evenson et al,.1998). The market price of germplasm captures mainly the so called use value of genetic resources (i.e. the value associated with the direct and indirect benefits resulting from use of germplasm by farmers and breeders). This is largely influenced by the technologies available and market forces. Use of advanced technology will increase breeders demand for germplasm and thus raise its value and market price. In the case of agriculture it is simply impossible to offer a reasonable estimate for the contribution of genetic diversity to crop production. Although genetic resources are the building blocks or raw materials for crop improvement programme and the importance of biodiversity for agriculture is sufficiently high to support broader conservation, the economic valuation of genetic diversity in breeding programme has been limited but it is necessary to prove that investment in conservation is worthwhile Earlier, genetic resources were considered priceless and it was considered that all genetic material has potential value. Valuation of biodiversity is a complex issue and has been done mainly by way of assessing its value for pharmaceutical and agriculture development but there are many other important valuation issues beyond the scope of these two. A measure of value for maintenance of genetic diversity through *ex situ* collection or for use of a specific type of biodiversity in plant breeding can be obtained but these values might change in future, even then, it is difficult to quantify the benefits of each and every accession held in the genebank.

The economic valuation of genetic resources is required to tackle the -

- Issues of sharing of benefits between providers and users of genetic materials at national and global level.
- Valuation is important to conservation policy makers.
- Property rights provide incentives for collection and preservation and endow plant genetic resources with value.

Scientists now agree that genetic resources have value and are convinced of collection and preservation of useful species such as rice, wheat, chickpea etc. and their wild relatives. The point of contention is how many species to conserve and how far to extend conservation efforts. While some (Wilson, 1988) argue that the potential value of genetic materials, combined with inherent uncertainty about the future, justify preserving essentially all known species. But the cost of protecting all of the world's genetic resources would be prohibitive. We should work out which species merit conservation. Plant breeders have traditionally drawn on intra- species diversity to improve crop yields, raise productivity and fight diseases and pests. Non agriculturists assume that protecting interspecies diversity is most urgent and equate 'biodiversity crisis' in terms of rapid loss of species and advocate preservation of species and ecosystem diversity. The cost of conserving germplasm thus necessitates some implicit ranking of the value of different species. This value must be based on current and future consumption and production values. A common justification for preserving germplasm is the insurance against potential outbreak of diseases and pests and role of genetic uniformity in susceptibility of crops to massive failure. Some economists arrive at rough estimates of value of *ex situ* collection by linking them to prevention of crop losses at a national or global level, for example, the 1970 epidemic of Southern corn leaf blight caused crop losses estimated to be worth 2 billion dollars, implying that the genetic improvement efforts that defeated the disease could be valued at the same amount. Although the extent of crop losses due to a disease does not necessarily indicate the true value of genetic materials.

Several studies have been conducted to place economic value on biodiversity for agriculture. As genetic diversity still has a functional ecological role in cropping systems, conservation and use are integrated. The value of genetic resources and biodiversity in general, reflects the increased well being that people derive from them – whether directly or through their use in production. The information contained in the ecological interactions among living organisms might be more important than the information contained in their genes. The loss of any species can disturb the delicate ecological balance of a natural system. This in turn can cause damaging effects for human beings. The extinction of a species for which many close substitutes exist matters much less than the extinction of a species with no close substitute. The importance of genetic diversity in livelihood strategies of small subsistence farmers seem to be undervalued in institutional plant breeding. Links have been established between accessible *ex-situ* collections and trait content and varietal production, because of the value of rare traits the rarest genetic resources may have the highest value, and wild species also have value. Collections become more valuable to breeders when they are evaluated.

Unfortunately, the economic benefits associated with the conservation and sustainable use of genetic resources for food and agriculture are poorly understood because of the complexity of the issue. The natural value of plant genetic resources is the value of the final product (e.g. a new variety) minus the value added by each activity involved in developing the finished product. Naturally occurring unknown and incompletely known plant genetic resources have option values. The option value reflects the future economic benefit to the society from saving and maintaining a sufficiently large biodiversity for future needs of society, in other words, minimizing actions which would limit the options for breeders in the future. Indirect production value that species can add from their services to the ecosystem. Future non consumption use value that is derived from preserving genetic resources as a form of insurance against an uncertain future.

Use of Plant-biodiversity

Direct

- Fundamental to basic need of food, fibre (clothing), medicine, forage etc.
- Infrastructure development (houding and construction)
- Rituals, Tourism, recreation and sustainable management of enviroment
- Economical-bioprospecting

Indirect

- Genes (diversity) for crop improvement to incerase and sustain food productivity
- Ecological balance and environment protection
- Material for scientific research, education and social welfare

Box : 1

Fast Facts on Loss of Biodiversity

International

- 50,000 to 2,50,000 of Plants and 800 vertebrates will be lost by 2020 AD (Raven 1987)
- IUCN estimates loss of 25% of world species by 2050 AD
- 10000 Chinese wheat varieties popular in 1949 reduced to 1000 only
- 7100 fruit varieties are growing only 20% of the corn·varieties todays
- Mexican farmers are growing only 20% of the corn varieties cultivated in 1930s
- In Andean countries many wild relatives and local races are under erosion

Box : 2

Fas Facts on Loss of Biodiversity

National

- IUCN list 17 Indian species in "Extinct Plant Species of the World" (few medicinal species)
- Red Data Book of BSI list around 1500 plant species as endangered (major component is medicinal species)
- About recently 30,000 rice varieties were grown in which are frared to reduce to 50 by 2000 (envfor.nic.in)

Box : 3

•••

2

Conservation Needs and Strategies

There is a global recognition that biodiversity at all levels- gene pools, species and biotic communities is important for many reasons and that it is being rapidly diminished by habitat destruction and other damaging influences resulting from human population growth, pollution and economic expansion. It is also now widely realized that the genes, species, ecosystems and the associated traditional knowledge and wisdom are being lost at an increasingly accelerated pace. Habitat destruction, genetic homogeneity in farming systems, and alien species invasion are some of the causes of erosion. The term "genetic erosion" first used by Harlan in 1972 signifies a dramatic shift in population structure within a crop species that can result from a range of natural or human-led processes and has become synonymous with the displacement of landraces by modern varieties. In the changing international scenario, global climatic changes, unsustainability of high input agriculture, rehabilitation of wastelands and problem soils and the search for novel genes for nutrition, biotic and abiotic stresses, diversification, environmental security have increased the focus on the extent of variability in genetic resources collections of crops, and the effectiveness of its utilization in raising the productivity through widening of genetic base and mapping genes for biotic and abiotic stresses.

Management, conservation and sustainable use of plant genetic resources are fundamental to ecologically sustainable development and food security of a nation. Loss of genetic diversity has serious implications on economic and social development of any nation.Three main processes cause loss of genetic diversity of cultivated crop species.

Genetic erosion : due to increasing uniformity in farming systems, overexploitation of useful genetic resources, (such as the plant species of medicinal value) without sustainable maintenance , introduction of exotic species etc. Although the direct effect of intellectual product on genetic erosion might be tenuous the effects can be very significant. Intellectual property enhances incentives for commercial plant breeding i.e. development of varieties with the largest market potential, crops with less commercial potential that are adapted to specific environmental niches or are better suited to the needs of small scale farmers risk being neglected and gradually abandoned. The impact of genetic erosion is felt differently by different research systems; for institutional innovation system impact is felt when breeding material is not readily available in genebanks . For community innovation system the loss is felt when breeding stock is no longer found in the field or local market.

Genetic vulnerability : It is due to intensive monoculture and narrow genetic base of crops. Global food security and rural incomes are at risk due to their excessive dependence on declining number of plant species. Diversified cropping systems such as those based on mixed cropping or agro-forestry are stable and conserve natural resources. Practices such as inter-cropping and poly culture that help to increase genetic diversity are being abandoned. Decline in number of varieties under cultivation is resulting in loss of valuable sources of nutritional traits in many crops. These trends also cause loss of local traditional knowledge and inequitable control of resources.

Genetic wipeout : It is mainly due to destruction of habitats and forests caused by man or natural calamities.

In the first World Conservation Strategy conservation is defined as "the management of human use of the biosphere so that it may yield the greatest sustainable benefit to present generations, while maintaining its potential to meet the needs and aspirations of future generations. Thus conservation is positive, embracing preservation, maintenance, sustainable utilization, restoration and enhancement of natural environment. The conservation of biological diversity is sought mainly for ethical or aesthetic reasons; the possibility of deriving direct economical benefits from it; and the protection from extinction of ecosystems and plant, animal and microorganism species. Conservation of biodiversity represents investments in products of potential utilization, while preservation of genetic resources emphasizes the need to characterize and use products of immediate importance. It is costly to preserve genetic materials whether *in situ* or *ex situ* and while we cannot expect conservation to yield windfall rewards in immediate future, the benefits of genetic conservation are long term.

Germplasm conservation provides multiple benefits-

- Increase productivity and food security (and consequently economic returns)
- Reduce pressure of agriculture on fragile areas, forests and endangered species
- Build stability, robustness, and sustainability of farming systems
- Contribute to sound pest and disease management
- Diversify products and income opportunities from farms
- Help maximize effective use of resources and environment (restore ecological health)
- Reduce dependency on external inputs
- Increase nutritional value and provide source of pharmaceutical products

Maintenance and conservation strategies depend on the kind of targeted diversity besides other factors, for example, ecosystem diversity with wild species is best conserved by *in situ* methods, species diversity in cultivated plants in gene banks and on farm, and intra-specific diversity in crop plants are best maintained in gene banks. Management of crop genetic resources includes their conservation or maintenance in a particular state, and the sustainable utilization of germplasm to meet the overriding need to increase crop intensity to meet future demands for food.

Until the 1980s, conservation of agricultural biodiversity was officially through collection of seed samples from farmer's fields and their storage in national and international gene banks. As a result, more than 6 million samples are stored worldwide in about 1500 gene banks. This response is now recognized to be a limited but important security backup strategy so long as these seed samples can be kept viable and safe, away from genetic contamination such as genetically modified organisms (GMOs), especially, in centers of origin and diversity of the world's crop. The Convention on Biological Diversity (CBD) ratified in 1992 has highlighted the need to employ *in situ* conservation as a complementary approach to *ex situ* conservation of plant genetic resources which shall need to be judiciously integrated so that the efforts of conservation are effectively realized for sustainable release of useful variability.

World's Major Seed Gene Bank
National Genebanks'Holdings

Gene bank	**Holdings**
India	354922
Germany	57068
Japan	105096
Korea Rep.	115837
Canada	212061
Brazil	232216
Italy	82057

Source: FAO Report, 2006

Conservation strategies - Options available

The Agro-biodiversity include species from very diverse groups, including species from different crop groups, habit and habitats with diverse reproductive biology. Thus for conserving such diverse plant biodiversity, a comprehensive and complimentary approach has to be followed which can maximise the benefit of different approaches to ensure optimum conservation of species or a gene-pool. Based on the threat, size of sample, nature and breeding behaviour of the species and the objectives of conservation, the conservation strategies have broadly classified into two categories as

a) *In situ* conservation which refers to conservation of species or genetic diversity within their ecosystem and natural habitat.

b) *Ex situ* conservation where the different components of the plant diversity are conserved outside their natural habitat.

In-situ Conservation

In situ conservation is concerned with maintaining species populations in the natural habitats where they occur, whether as uncultivated plant communities or in farmer's fields as part of existing agro-ecosystems. In the Global Action Plan of FAO (FAO, 1996c) it has now been recommended to accord priority for *in situ* methods of maintenance. Thus this method of maintenance is now being discussed widely and has achieved some success but is still based on weak scientific status. In fact most current considerations and experiments with on farm conservation deal with few species. *In situ* conservation of crop plants involves the conservation on-farm of local crop cultivars (or landraces) with the active participation of farmers.

In situ conservation has the potential to:

- Conserve the evolutionary process of local adaptation of crops to their environment.
- Conserve diversity at all levels of ecosystem, species and genetic diversity within species and act as backup to gene bank collections.
- Conserve ecosystem services critical to functioning of earth's life support system.
- Improve livelihoods for resource poor farmers through economic and social developments.
- Maintain or increase farmers control over their crop genetic resources.
- Ensuring that farmer's effort are an integral part of national plant genetic resources system and involve farmers directly in developing options for adding benefits of local crop diversity.

Before proceeding to any in situ programme it is advisable to access the diversity and importance of the species occurring in the area which is to be demarcated for in situ conservation. Generally the species which are endemic and indigenous in the country(ies) are recommended to be entirely or partly conserved *in situ* in the following cases:

A. *The species comprise cultivated as well as wild growing forms if*

- The species is difficult or expensive to manage by ex-situ methods;

- The species is very widely distributed; Many important herbage plants are included in this cathegory. Such a conservation approach may provide a less expensive protection than *ex situ*, particularly as extensive collecting and multiplication is a heavy burden.
- The species is endangered and it will become extinct if measures are not taken.

B. *The species comprise only wild forms in the area but is cultivated somewhere else*

By international undertakings, such as the CBD and the FAOs Undertaking for the conservation of genetic resources, the Genebank may be tied to the international network of gene banks. However, collecting, conserving and multiplying even species of little or no interest for the country(ies) for which the Genebank has responsibility may be too costly to manage. To meet the demand from other parts of the network, these species can be conserved *in situ*

C. *The species is nowhere cultivated (wild relative of cultivated species)*

Wild species (including wild forms of cultivated species) with a functional mechanism of seed dispersal are in general more complicated to handle *ex situ*, than cultivated plants. Therefore *in situ* conservation, being less costly, can be recommended for wild species.

Action Plan for In situ Conservation

The major activities which are required for in situ activities for crop germplasm conservation, are as follows.

(1) Initial surveying

- Taxonomic clarification
- Determination of distribution of species
- Correlation of ecological and geographical factors
- Determination of genotypic variability by growing out of samples or other experimental methods.*

(2) Formulation of conservation objectives

- Determination of short- and long-term needs for the germplasm of the wild species of the targeted genepool

- Determination of the extent of conservation of alleles which is necessary e.g. all alleles (including the most rare), all alleles with greater than a 5% frequency etc.
- Determination of a satisfactory level of security for maintenance of alleles
- Determination of minimum level of access to target germ-plasm as related to convenience, dependability and costs for procurement.

(3) *Determination of minimum requirements for conservation of targeted species (based on conservation objectives) such as*

- Number of populations
- Size of each population
- Ecological and environmental conditions
- Size of habitat units

(4) *Assessment and choice of sites in terms of minimum requirements for conservation including*

- Existing, protected natural areas
- Non-allocated wildlands
- Lightly exploited sites which could be acquired or managed.

(5) *Reserve design*

- Integration of minimum requirements for preservation of targeted populations into larger conservation and development planning units
- Design of reserve boundaries possibly in clusters or as multiple reserves
- Zoning and related regulation for "mitigation" of potentially adverse impacts of land use activities in adjacent areas which could influence targeted populations within reserves.

(6) *Reserve management*

- Formulation of "prescriptions" to off-set any negative ecological impacts from fragmentation

- Formulation of regulations and prescriptions to control human activities and to off-set negative impacts which are produced by them
- Implementation of prescriptions along with on-going evaluation and revision of prescriptions.

(7) Monitoring

- On-going research on targeted populations and genetic variability
- On-going assessment of the biological requirements of the targeted species and the conditions of respective populations
- On-going assessment of human activities and related impacts as they affect targeted species.

(8) Access to germplasm

- Regulation of procurement of genetic material within reserves so as to maintain minimum levels of conservation
- Documentation such as mapping, organization development and training guides.

Ex Situ Conservation

Ex situ conservation is the conservation and maintenance of samples of living organisms outside their natural habitat, in the form of whole plants, seed, pollen, vegetative propagules, and tissue or cell cultures. Ex-situ conservation methods are well suited to capture and store all alleles/ genotypes safely on long term basis, while in-situ conservation permits evolution to continue and populations to adopt changing environment. This can further be achieved through different man-made conservation methods that includes

I) Field gene banks

II) Seed gene banks

III) In vitro gene banks

IV) Cryo-gene banks, and

V) DNA gene banks

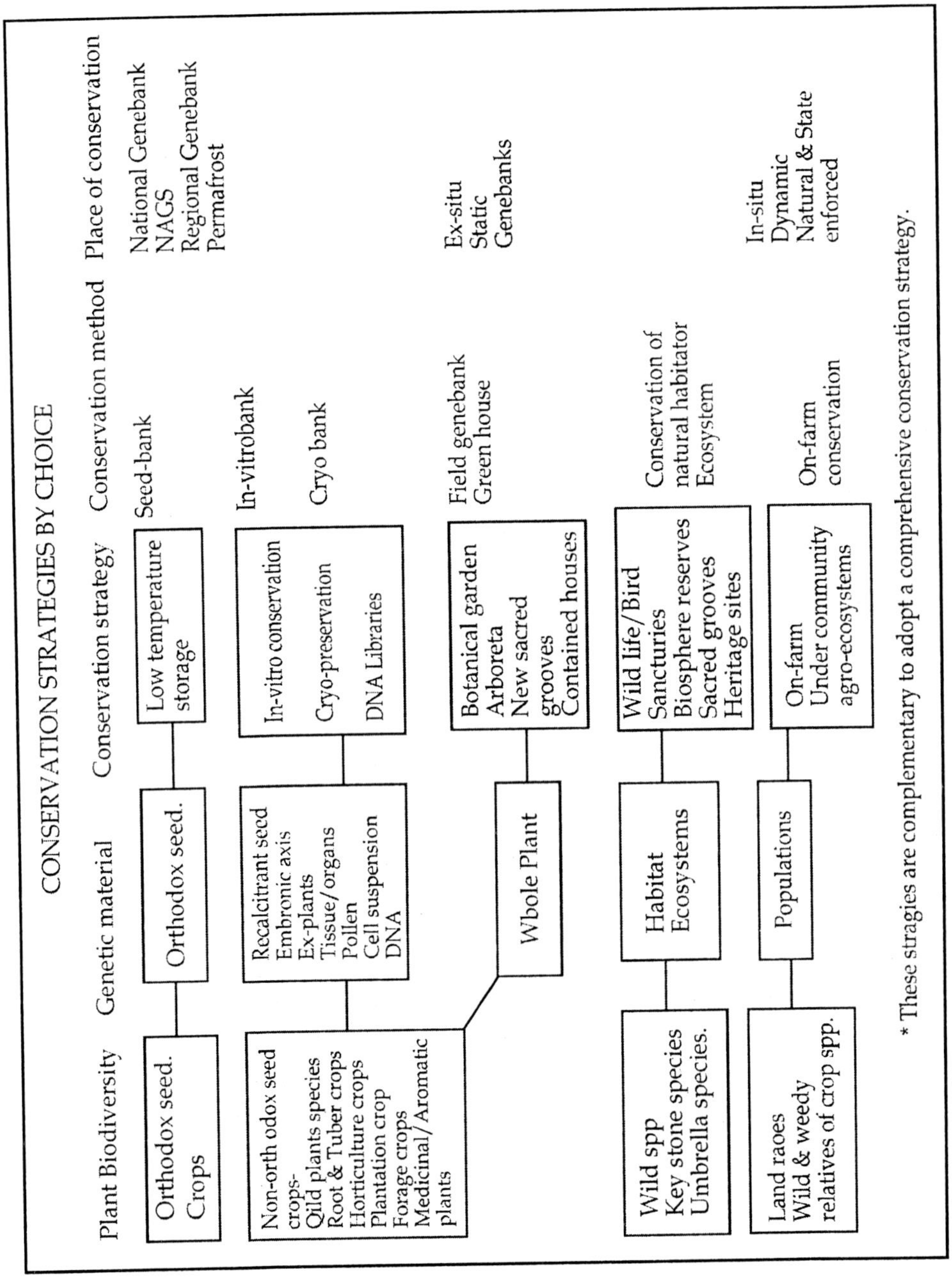
CONSERVATION STRATEGIES BY CHOICE
Plant Biodiversity
Genetic material
Conservation strategy
Conservation method
Place of conservation
Orthodox seed. Crops
Orthodox seed.
Low temperature storage
Seed-bank
National Genebank
NAGS
Regional Genebank
Permafrost
Non-orth odox seed crops-
Qild plants species
Root & Tuber crops
Horticulture crops
Plantation crop
Forage crops
Medicinal/Aromatic plants
Recalcitrant seed
Embronic axis
Ex-plants
Tissue/organs
Pollen
Cell suspension
DNA
In-vitro conservation
Cryo-preservation
DNA Libraries
In-vitrobank
Cryo bank
Wbole Plant
Botanical garden
Arboreta
New sacred grooves
Contained houses
Field genebank
Green house
Ex-situ
Static
Genebanks
Wild spp
Key stone species
Umbrella species.
Habitat
Ecosystems
Wild life/Bird
Sancturies
Biosphere reserves
Sacred grooves
Heritage sites
Conservation of natural habitator
Ecosystem
Land raoes
Wild & weedy relatives of crop spp.
Populations
On-farm
Under community
agro-ecosystems
On-farm conservation
In-situ
Dynamic
Natural & State enforced
* These stragies are complementary to adopt a comprehensive conservation strategy.

Ex situ conservation has several purposes:

- Rescue threatened germplasm.
- Produce material for conservation biology research.
- Bulk up germplasm for storage in various forms of ex situ facility.
- Supply material for various purposes to remove or reduce pressure from wild collecting.
- Grow those species with recalcitrant seeds that cannot be maintained in a seed store.
- Make available material for conservation education and display.
- Produce material for reintroduction, reinforcement, habitat restoration and management.

The important features of an *ex situ* conservation programme for any particular species are:

- To be an important backup measure other than *in situ* conservation
- To ensure that a wide range of the diversity (phenotypic and genotypic) available in a species is conserved
- To manage the regeneration of the species outside its original natural range (provenance) in a more controlled way (which is likely to further develop the population(s) for use or conservation).

Levels of the genetic hierarchy and their relation to *ex situ* conservation

- *Genes:* Important functional genes—for disease resistance, for example—could be mapped using molecular markers. With new molecular technologies, the gene could be sequenced to examine the possible molecular basis of the resistance, and even cloned and introduced in the same or other species. However, conservation of such genes typically means conserving the actual genotypes in which they occurred.
- *Genotypes:* These are individuals that appear to be representative of the population, or for some trait of interest (either phenotypically or from clonally). In order to maintain the

genotype, material must be captured by vegetative reproduction. Collection of seed from an individual does not generally preserve the genotype of that individual.

- *Populations (provenances):* Although we are usually interested in conserving genetic variation which is representative of populations, we are ultimately required to sample a sufficient number of individuals that represent the gene frequencies in that population. The question then arises of how to allocate sampling resources such as numbers of trees, type of tissue, e.g. seed, cost of sampling, within and among populations.
- *Ecosystems:* We cannot measure the genetic variation of an ecosystem, because intraspecific genetic variation is the property of a species, and representative populations may transcend several different types of ecosystems (dry or moist, warm or cool, assemblages and competition with different species, etc.). The only way we can say we are 'sampling ecosystems' for genetic variation is to sample populations, in an attempt to capture genetic variation that has developed among populations as a result of historical colonization events, different selection pressures, or gene flow.

Seed Genebanks

Seed Genebanks are an ex situ conservation strategy designed to capture and conserve genetic diversity within and among species. In genebanks, germplasm is placed in suspended animation so that desirable allelic combinations and rare alleles of a species are available for the future. Plant germplasm is really a collection of propagules: seeds or pollen to preserve the genetic composition of populations, and cuttings, buds, rhizomes or cell cultures to preserve specific genetic combinations of individuals. Most genebanks have a completely utilitarian goal: plant breeders use collections to make higher-yielding, more resistant crops and ecologists preserve threatened populations until they can be reintroduced into restored habitats. The seed gene banks as a tool to conserve the germplasm has many advantages as

i) They provide immediate access to plant species allowing researchers and conservationists to evaluate them for properties such as now sources of medicine, nutrition and genes for wild traits,

ii) To generate information that could provide helpful information for conserving the remaining population in the nature

iii) Germplasm stored away from the natural calamities can be used to reinstate the species into the suitable habitats if the need arises.

The collections at Seed Genebank are maintained in three different forms as

a) *Base Collections* where fresh viable pure seeds after drying to 3-7 percent moisture are kept in hermatically sealed tri-layered aluminium foil pouches at – 20°C in long term modules. The base collections are not routinely disturbed except for monitoring the seed viability and to replenish the active collections, in case they are lost.

b) *Active Collections* where the seeds are stored in modules at 5°C and the relative humidity of 35-40 percent in various containers such as cloth-bags, metal cans or glass jars. These collections are maintained at different National Active Germplasm Sites (NAGS) for distribution and regular use.

c) *Working Collections* where seeds are stored for two or three seasons at ambient temperatures of 20-25°C and relative humidity of approximately 35-40 percent. The breeder's seeds are best stored under these conditions.

Action Plan for Seed gene bank

The seed gene banks are most effective and economic form of the ex situ conservation for the plant species that produce orthodox type of seed i.e., seeds which do not loose their viability even if dried to low moisture content . They, by virtue of small size, occupy less space and represent wide genetic variability as each seed has unique genetic information.

- Augmentation of the seed germplasm
- Registration of the sample by allotting national identity
- Testing of moisture content using oven dry methods
- Testing seed viability (ISTA germination standards)
- Estimating viability using Quick viability test (TTC Test)

- Drying of the seeds to low moisture levels (5-7%) using dry seed cabinets
- Packaging of the seeds in hermetically sealed tri-layered aluminum foil pouches
- Data base management of the stored accessions.
- Storage of these samples at -18 ^{0}C in long term modules

In-vitro gene banks involves plant tissue culture technique, also called micro-propagation. It is a practice used to propagate plants under sterile conditions, often to produce clones of a plant. It is widely used in forestry and in floriculture. Plant tissue culture can also be used to conserve rare or endangered plant species as In vitro Repository for difficult germplasm. Different techniques in plant tissue culture may offer certain advantages over traditional seed conservation methods which includes

- The production of exact copies of plants that produce particularly good flowers, fruits, or have other desirable traits.
- To quickly produce mature plants.
- The production of multiples of plants in the absence of seeds or necessary pollinators.
- The regeneration of whole plants from plant cells that have been genetically modified.
- The production of plants in sterile containers that allows them to be transferred with greatly reduced chances of transmitting diseases, pests, and pathogens.
- The production of plants from seeds that otherwise have very low chances of germinating and growing,
- To clean particular plant of viral and other infections and to quickly multiply these plants as 'cleaned stock' for horticulture and agriculture.

Action Plan for *In vitro* gene banks

Modern plant tissue culture is performed under aseptic conditions under filtered air. Living plant materials from the environment are naturally contaminated on their surfaces (and sometimes interiors) with microorganisms, so surface sterlization of starting materials or explants in chemical solutions (usually alcohol or bleach) is required. Mercuric chloride is seldom used as a plant sterilant today, as it is dangerous to use, and is difficult to dispose of. Explants are then usually placed on the surface of a solid culture medium, but are sometimes placed directly into a liquid medium, particularly when cell suspension cultures are desired. Solid and liquid media are generally composed of inorganic salts plus a few organic nutrients, vitamins and plant hormones. Solid media are prepared from liquid media with the addition of a gelling agent, usually purified agar.

The culture environment

When cultured *in vitro*, all the needs, both chemical and physical, of the plant cells have to met by the culture vessel, the growth medium and the external environment (light, temperature, etc.). The growth medium has to supply all the essential mineral ions required for growth and development. In many cases (as the biosynthetic capability of cells cultured *in vitro* may not replicate that of the parent plant), it must also supply additional organic supplements such as amino acids and vitamins. Many plant cell cultures, as they are not photosynthetic, also require the addition of a fixed carbon source in the form of a sugar (most often sucrose). One other vital component that must also be supplied is water, the principal biological solvent. Physical factors, such as temperature, pH, the gaseous environment, light (quality and duration) and osmotic pressure, also have to be maintained within acceptable limits.

Plant cell culture media

Culture media used for the *in vitro* cultivation of plant cells are composed of three basic components:

1. Essential elements, or mineral ions, supplied as a complex mixture of salts;
2. An organic supplement supplying vitamins and/or amino acids; and
3. A source of fixed carbon; usually supplied as the sugar sucrose

Plant growth regulators

We have already briefly considered the concepts of plasticity and totipotency. The essential point as far as plant cell culture is concerned is that, due to this plasticity and totipotency, specific media manipulations can be used to direct the development of plant cells in culture. Plant growth regulators are the critical media components in determining the developmental pathway of the plant cells. The plant growth regulators used most commonly are plant hormones or their synthetic analogues.There are five main classes of plant growth regulator used in plant cell culture,Viz., auxins, cytokinins, gibberellins, abscisic acid and ethylene.

Culture types

Cultures are generally initiated from sterile pieces of a whole plant. These pieces are termed 'explants', and may consist of pieces of organs, such as leaves or roots, or may be specific cell types, such as pollen or endosperm. Many features of the explant are known to affect the efficiency of culture initiation. Generally, younger, more rapidly growing tissue (or tissue at an early stage of development) is most effective. Several different culture types are Callus, Cell-suspension cultures, Protoplasts, Root cultures, Shoot tip and meristem culture, Embryo culture and Microspore culture

Cryo-banks

Not all seeds are "orthodox", however. Some seeds are killed by drying, and these are known as "recalcitrant" seeds. Many of these seeds are from plants that grow in environments that are constantly moist and which germinate soon after they are shed from the plant. In other cases, the plant may make very few or no seeds. Because seed freezing is not possible in these cases, tissue freezing can be used. Various methods for cryo preserving tissues have been used are encapsulation dehydration method, slow freezing (two-step freezing) and encapsulation vitrification method. Cryo preservation is a collection of methods for freezing living tissues at very low temperatures, usually in liquid nitrogen (LN) (-196oC/-320oF). At these temperatures, living tissues are stable and may be preserved for decades or longer. There are many techniques for protecting living tissues at these very low temperatures and ensuring that they will resume growth when they are thawed (Fabre and Dereuddre, 1990, Cryo-Letters 11: 413-426).

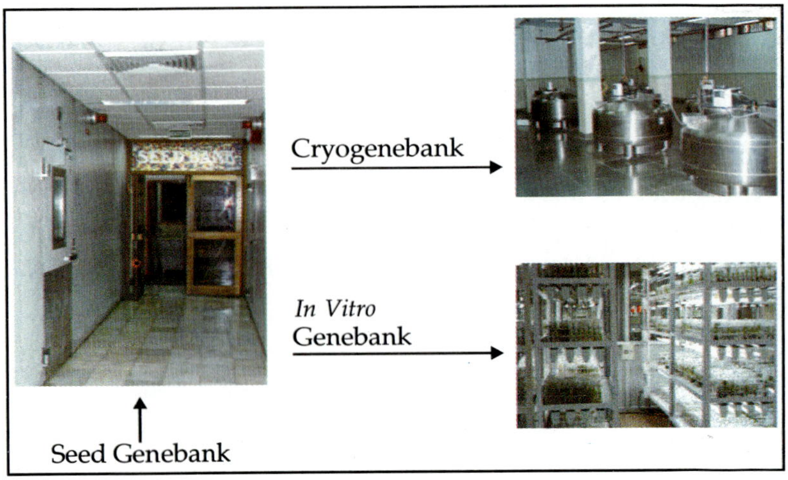

Germplasm Conservation

- Shoot tips from tissue cultures are dissected from the shoots and used for freezing. Only the top 1 mm of the shoot is used. One method for cryopreserving shoot tips is the encapsulation dehydration method.
- The tiny shoot tips are placed into a very viscous solution of alginate
- The alginate with the shoot tips is drawn up into a pipet. The alginate is added dropwise to a solution of calcium chloride, which causes the alginate to gel. These drops become "beads" of gel around the shoot tips
- The beads are placed into a liquid medium with a high concentration of sucrose for 18 hours overnight on a rotary shaker. The sugar acts as a "cryoprotectant", pulling water out of the tissues and replacing the water in the tissues.
- Then, the beads are placed on sterile filter paper under the air flow of the laminar flow hood to dry for 4 hours. Once they are dried, the beads are transferred into sterile "cryovials".
- These are plunged into liquid nitrogen, preserving the tissues.
- The tissues are thawed by placing the cryovial on the benchtop for 20 minutes and then transferring the dried beads to tissue culture medium. There they will rehydrate and the shoot tips inside the beads will resume growth.

•••

3

Global Warming *vs* Biodiversity Conservation

Global climatic changes continue to be the challenging issues among the scientific and societal communities, whether they are the consequences of ozone depletion or environmental degradation or water pollution or land desertification. Some of the predicted deleterious climatic changes expected by the end of twenty-first century are

- Rise in the global mean temperature (1-3.5° C)
- Rise in the global mean sea level (approx. 50 cms)
- Impropriated changes in the day and night durations
- Unpredicted floods or drought
- Frequent incidence of cyclones or hurricanes or tsunamis

All these climatic changes have an intrinsic adverse effect on the prevailing biodiversity in general and on agro-biodiversity in particular. Instability in the local climate of a region greatly affects the agro-ecosystems and thereby influencing the genetic and species diversity. The utmost concern of the present day situation is to develop strategies to minimize the after effects of these climatic changes in short and long run through policy and technological options.

The current rate of biodiversity depletion is estimated to be hundred fold higher than the recent past may because of the large-scale habitat destruction. Recent attack of tsunami in the coastal areas of Thailand, Indonesia region is the eyewitness to this destruction of natural populations is still continuing there have been increasing efforts to stockpile and conserve wild medicinal plants in particular in the centralized seed banks gene banks are the most convenient facility to conserve the germplasm in safest mode.

Preparing for climate change is now an important environmental issue worldwide. Warming of climate is now evident from observations of increases in global average air and ocean temperatures, widespread melting of snow and ice and rising global average sea levels and it is likely that further changes will result from increasing concentration of CO_2 and other radiatively active gases in atmosphere through greenhouse effect. Annual greenhouse gas emissions from human activities have risen by 70% since 1970. Significant changes of climate are likely to occur over the next 50 years. Present assessments indicate that increases in global mean temperature of between 1.5°C and 6.4 °C and sea levels from 18 cm and 59cm are likely to occur during the century. Combustion of fossil fuel, continuing deforestation and increasing methane emissions from ruminant livestock and several other factors have led to increase in CO_2 levels in recent decades. CO_2 levels have been rising by 0.53% or 2 ppm per year since 1958.Climate modelers consider that among other environmental changes the atmospheric level of CO_2 is expected to double by the middle of the next century (i.e. twice the level of 280 ppm). There is still however, much uncertainty surrounding the timing and magnitude of greenhouse warming. From 1976 to 2004, when the world's average surface temperature rose 0.49°C, global level atmospheric water vapour rose 2.2 percent. By 2100 humidity levels could increase by another 10 percent.

The Arctic ice cap is melting even more rapidly than thought; by 2050, 40% of the ice cover in the Arctic ocean could be gone, a loss that wasn't suppose to happen for 100 years. It is also projected that by this time almost one fourth of the flora and fauna might disappear. Even slight changes in one environmental factor such as temperature would trigger changes in many other abiotic and biotic factors which in totality may constitute a major change in the adaptive conditions of a population. In addition to temperature and humidity other effects including potentially critical changes in regional rainfall pattern, evaporation, storms, droughts (in semiarid regions), run-off and soil moisture and heat stress are likely to occur. There is general agreement that climatic change will occur within a far shorter time span than ever experienced before, world temperature will rise and warming is predicted to be greatest at higher latitude, rainfall patterns are likely to be altered with some areas becoming wetter and others much drier than present conditions. Weather patterns may become even more variable than at present. The expected change will not only be global but also long lasting and may or may not have the same direction everywhere. Green house gases, climate change and ozone depletion also mean such unpredictable changes like expansion of desert and grasslands and decline in forests ecosystems, shifts in disease patterns etc. Since greenhouse gases linger for decades, despite remedial international controls on emission of green house gases implemented in near future the sea level rise would continue for long because of the momentum of warming even if greenhouse gas levels are stabilized. With it the risk of extreme weather events such as floods, droughts and heat waves. In addition, the effects of future emission means that we will have to cope with the green house effect and its consequences for agriculture and genetic resources also.

Climate change poses a threat to the continued survival of these genetic resources and is yet another potential cause of genetic erosion. The World Conservation Union (WCU) in its annual red list of threatened species warned of a "global extinction crisis" and predicts that more than 16, 000 species are at risk of extinction. Risks to unique and threatened systems, such as polar or high mountain ecosystems, coral reefs and small islands are more critical. These considerations call for priority of preserving adaptability over preserving the current genetic status of populations.

The preservation of ecosystem and a diversity of habitats and topography are important if plant species are to retain flexibility to respond to climate changes. The influence of altered habitat in the background of climate change on the wider gene pools of crops not immediately threatened and normally left in the wild need to be worked out. Since there are no precedents so analysis of the role of plant genetic resources must be through predictions. With global warming we shall be faced with the dual challenge of ensuring that genetic conservation takes place as well as utilization of germplasm to breed new varieties. In order to be able to respond to imminent changes in the global climatic pattern plant breeders will need sources of new genetic variation and breed new varieties to meet these challenges. Crop genetic resources of cultivated plants are mostly conserved because of their prospective economic value which includes resistance to pest and diseases In this context resources of new crops can be explored and exploited. The productive potential will be represented by the genomic constitution which produces the desired phenotypes under the modified environmental conditions. For agriculture, crop genetic resources activities should take account of changes in climate in terms of collection, conservation and their utilization.

The magnitude of projected climate change demands the immediate evaluation of technical and policy options. Some of the key concerns are - will climatic change affect the current production of stable food crops and forages, especially in relation to productive and marginal areas? Is gene conservation under global climate change meaningful? On the one hand is the fear of threat to food supplies by climate change and the unlikelihood of developing new crop strains quickly enough to keep pace with rapid changes in environmental conditions. On the other hand is optimistic viewpoint that agriculture has been often encountering climate change and crop genetic resources are capable of adapting to these stresses as in the past. The importance of plant genetic resources conservation can be easily appreciated in this context. The methodologies for genetic conservation in light of pending global climatic change also need a review. Both dynamic methods of germplasm conservation as well as static methods of conservation (such as in a genebank) are needed in order to overcome local and temporary deleterious environmental changes. In many areas of the world where traditional varieties can still be found introgressing and hybridizing with related wild species the threat of global warming will disrupt this process of dynamic gene flow. A better understanding of eco-geographic variation of both wild and cultivated species would be helpful The strategy employed for conservation of germplasm depends on degree of change as well as on nature of material required to be conserved. For this we should be able to identify the sort of genetic resources required to face these challenges. Evaluation and characterization of genetic resources conserved in the gene bank is now the most important activity and will help to identify the useful genes it contains to assist in development of crop varieties more appropriate for altered climates anticipated in future. Screening for drought tolerance, raised temperatures, salinity, and physiological effects of CO2 i.e., enhanced photosynthetic rates on water availability and on pest, diseases and their vectors and the critical tolerance ranges of crop with respect to climatic factors to identify those adapted to projected climates is needed.

Any change in mean climate would be manifest in a long-term spatial shift in agro climatic zone necessitating a change in focus of the breeding programmes, large latitudinal shifts would entail attention to photoperiodic adaptation. As plant breeding is a long- term enterprise the importance of timely and accurate predictions of changes in climate cannot be overemphasized. One of the most important implications of this would be for germplasm characterization and utilization. Accessions in germplasm collection are often evaluated at a site and then classified into categories such as early, medium, late, etc. However, such evaluation data are site specific. Crop improvement by effective utilization of the germplasm collection now available for the major world crops has been problematic mostly because of quantitative nature of gene expression. The characterization data required for genotypes may not be the same for breeders targeting different agro-ecological zones. Genotypes with reduced sensitivity of developmental rates to temperature would be needed to maintain more consistent flowering dates despite global warming. This seems feasible in many crops, for example, four-fold difference in temperature sensitivity is found in barley genotypes. Alternatively, genotypes which are intrinsically of duration too long for current use in cool climates could become more important and so need to be both conserved and characterized. In warmer climates more consideration should be given to increasing the value of optimum temperature and to lessening the deleterious consequences of exposure to supra optimal temperature regimes. While considering the effect of change in temperature the possible effects of photoperiod must also be taken into consideration. The main question then is how best to recognize, quantify, screen and predict the consequences of incorporating suitable combination of adaptive traits into breeding programmes. The need is to characterize accessions in germplasm collection with respect to sensitivity of phenology to photo thermal environment. With the increasing likelihood of significant climatic changes further research is now warranted on the quantitative response of crop plant development to supra optimal temperatures which have been comparatively under researched.

•••

4

Wild Relatives of Crop Plants (WRCPs) - Present Status and their Conservation

The cultivated plants used by the man today have been derived from the wild species through selection or hybridization for getting desirable characters in the progenies. During this process of selection or domestication a very small number of genes in a specific combination have been accumulated in the crop plants leaving behind a vast gene pool of untapped genes in wild species. The cultivated crop plants diversity can be broadened by using this gene pool. The wild species in primary gene pool includes wild progenitors of crops, wild and weedy forms, thereby producing fertile hybrids with cultivated species (C Harlan and de Wet,1971) whereas wild species of secondary and tertiary gene pool can only be used for some crops for a limited number of genetic traits (Harlan 1975). But, the science of genetic engineering has opened almost unlimited possibility for the conservation and use of this wild plant species wealth.

It is of prime importance to understand taxonomically, ecologically & genetically the species diversity and to sort out the intricate relationship of species, subspecies, varieties and ecotypes and to group these according to the degree of relationship in a given gene pool (Hoyt 1988). The wild gene pool includes the naturally occurring wild types, weedy types protected, semi-domesticated and domesticated diversity, where as the crop gene pool includes at the cultivars, wild relatives and related wild species containing all the gene available for breeding use individually each crop gene pool (GP) can be further classified as

1. *Primary Gene Pool (GP1)* : Where wild species contribute more readily fertile hybrids with easy gene transfers.

2. *Secondary Gene Pool (GP2)* : There include wild species with can be crossed with GP-1 species producing hybrids etc have (usually sterile) low fertility, rate. In this case gene transfer is possible but very difficult.

3. *Tertiary Gene Pool (GP3)* : The wild species included here can he crossed with other gene pools but hybrids produced on sterile, Gene transfer in possible only through genetic engineering or other radical means.

4. *Gene Pool created by human intervention* : In present scenario of techno-era or genetic engineering era, more distantly related genera are also contributing to this gene pool because of sophisticated techniques of recombinant DNA transfer or with the development of transgenic plants where even a animal or microbial gene can he transferred to the plant cells for the transfer of desirable trait from animal kingdom to plant of gene pool has widen across the boundaries of same taxa.

Thus, there is a strong interest in maintaining and documenting the total range of genetic diversity of WR cultivated plants in primary, secondary and tertiary gene pools. The present compilation includes all the wild progenitors (closely related to cultivated plants), and the wild species relatively or distantly related to cultivated genera or even unrelated taxa which can be utilized in the improvement of agricultural production and for maintaining sustainable agro-eco-system. The information on their occurrence is available from different sources like herbaria/floristic accounts/floras.

Arora and Nayar (1984) earlier published a comprehensive account of wild relatives of crop plants occurring in India, listing 334 species of wild relatives. The present estimated strength of these wild relatives is about 517 species. The probable increase in the number of species may be due to the selection pressure under the changing environment and habitat destruction alongeith the introduction of many alien species in the Indian sub-continent. Till date no studies have been made to assess the present status of wild relatives.

Realising the importance of these wild relatives, emphasis was given on their conservation and collection under the NATP (PB). Analysis of the information thus generated showed a large number of new additions in the reported list of Arora and Nayar (table 1). In total 183 new wild relatives of different crop plants (table 2) are now being occurring in different agro-climatic zones of India. Most of the newly reported wild relatives are in the crop group of medicinal plants, tuberous crops and fodder crops, the groups which were not taken into account by Arora and Nayar (1984).

Table 1: Crop Group-wise Comparative account of the WRCPs

Crop Group	No. of WRCPs	
	Present Status (2005)	As reported by Arora and Nayar (1984)
Cereal and Millets	59	51
Legumes	68	31
Vegetables	61	54
Tuberous types	**13**	–
Fruits	143	109
Oilseeds	20	12
Spices and Condiments	38	27
Medicinal Plants	**32**	–
Fibers	29	24
Fodders	**13**	–
Others	41	26
Total	**517**	**334**

Table 2: Species-wise details of the new additions of WRCPs.

Cereals and Millets	*Allium govanianum*
Amaranthus blitum	*Allium wallichi*
Digitaria duthieana	*Allium macranthum*
(Syn.Paspalum duthieana)	*Allium phariense*
Digitaria stewartiana	*Allium stracheyi*
(Syn.Paspalum stewartiana)	*Amomum ghaticum*
Oryza afficinalis	*Amomum microstephanum*
Legumes	*Trigonella podperae*
Vigna khandalensis	*Trichosanthes nervifolia*
Crotolaria barbata	***Oilseeds***
C.clarkei	*Madhuca bourdillonii*
C.clavata	*Madhuca diplostemon*
C.digitata	*Madhuca insignis*
C.filipes var trichocarpa	*Sesamum alatum*
C.fusonii var glabra	*S.mulayanum*
C.globosa	*Tuberous types*
C.grahamiana	*Dioscorea deltoidea*
C.kodaiensis	*Dioscorea rogersii*
C.madurensis	***Fruits***
C.lutencens	*Garcinia cadelliana*
C.longipes	*Garcinia imbertii*
C.meeboldii	*Garcinia kingii*
C.noveoides	*G.talbotii*
C.penduncularis	*G.wightii*
C.priestleyoides	*Phoenix rupicola*
C.rigida	*Prunus pygeoides*
C.sandoorensis	*Prunus cornuta var integrifolia*
C.scabra	*P.himalaica*
C.stocksii	*P.glauciphylla*
C.subperfoliata	*P.rufa var imanishii*
C.tecta	*P.hargraonensis*
C.tetragonoloba	*P.wattii*
Rhynchosia beddomei	*Capparis cinerea*
Rhynchosia velutiva	*C.diverisifolia*
Vegetables	*C.fusifera*
Ipomoea clarkei	*C.padryphylla*
Lactuca benthamii	*C.rheedii*
Lactuca filicina	*C.shevaroyensis*
Lactuca cooperi	*Syzyium andamanicum*
Lactuca undulata	*S.beddomei*
Allium gilgiticum	*S.benthamianum*
Allium caesioides	*S.bourdillonii*

table contd...

table contd...

S.courtalleuse	*A.nallamalayana*
S.gambleanum	*Andrographis rothii*
Vanilla wightiana	*Asparagus jacquemonti*
Fodder grass	*A.rottleri*
Bracharia nilagirica	*Astragalus aegacanthoides*
Chloris wightiana	*A.bakeri*
Dichanthium armatum	*Astragalus heydei*
Dichanthium compressum	*A.orothrephes*
D.grahamii	*A.oxyodon*
D.maccannii	*A.zemuensis*
D.pandraganieuse	*Eremostachys superba*
D.woodrowii	*Pueraria bella*
Eragrostis rottleri	*Aconitum ferox*
Hedysarum astragaloides	***Miscellaneous***
Hedysarum cachemirianum	*Acacia camphellii*
H.microcalyx	*A.hohenachkeri*
Moghania gracilis	*A.wightii*
(Syn. Flemingia gracilis)	*Grewia gamblei*
Spices and condiments	*G.pandaica*
Piper berberi, P.argyrophyllum,	*Cryptocarya ferrarsii*
P.hapnium, P.altenauatum,	*E.lawsonii*
P.galeatum, P.beddomei	*C.neilgherrensis*
Cinnamomum riparium	*Dalbergia congesta*
Cinnamomum travancoricus	*Rosa bracteata*
Eugenia argentea	*R.clinophylla var clinophylla*
Eugenia discifera	*R.clinophylla var glabra*
Eugenia singampattiana	*R.clinophylla var parviflora*
Medicinal and Aromatic plants	*Vanilla andamanica*
Andrographis beddomei	*Vanilla walkeriae*

•••

5

The Seed and its Structure

After fertilization a series of changes takes place in the ovule and as a result seed is formed. Ovules are structures of seed plants containing the female gametophyte with the egg cell, all being surrounded by the nucellus and 1-2 integuments. In angiosperms the double fertilization results in formation of the diploid embryo and the triploid endosperm. Thus a fertilized mature ovule is called a seed.

Development of the Embryo : After fertilization the egg-cell or ovum secretes a cellulose wall round itself and becomes the osspore. The latter divides into two cells – an upper (away from the micropyle) and a lower (towards the micropyle). The lower one lying towards the micropyle further divides in one direction into a row of cells, called the suspensor. The suspensor, as it elongates, pushes the developing embryo deep into the embryo-sac, and it also acts as a feeding organ to the embryo during the formation of the latter. For this purpose the basal cell of the suspensor often enlarges and acts as an absorbing organ. The suspensor, however, becomes disorganized as the radicle is formed. The terminal cell of the suspensor lying text to the embryonal mass is called the hypophysis cell (hypo, below; physis; growth). It divides and gives rise to the apex of the radicle. The upper cell enlarges and divides by three walls at right angles into octant (eight cells), four cells lying towards the suspensor forming the posterior octants and the other four cells lying away forming the anterior octants. Each octant then divides by a wall parallel to its curved surface differentiating into a central norms. Thus, a surface (superficial) layer of cells and a central mass of cells, the latter being. The embryonal mass gives rise to the whole of the embryo except the root-tip. The cells of the embryonal mass divides repeatedly and the various sparts of the embryo become differentiated.

In monocots embryo: Although the mono-cotyledonous embryo follows the same general trend of development as the dictyledonous one of the former has only one cotyledon against the latter's two. Besides, in some monocotyledons the cotyledon is terminal is the case as in dicotyledons. There is also a considerable variation in the details of development of monocotyledonous embryos. A typical case may be as follows. The osspore divides into two cells. The upper one (lying towards the micropyle) enlarges considerably and forms a massive suspensor. The lower cell is the embryonal cell. It soon divides into two. The terminal cell of these two by repeated divisions in different planes gives rise to the single cotyledon. The other cell also divides in the same way and give rise to the stem-apex, the hypocotyls and the root-tip.

Development of the Endosperm: In all families of flowering plants with the exception of *Orchidaceae* and *Podostemaceae* as a result of triple fusion. The definitive nucleus, now called the endosperm nucleus, tends to grown into a food storage tissue called the endosperm, usually triploid (*3n*), some-times tetraploid (*4n*). In due course the endosperm may be a permanent feature of the seed (as in all endospermic seeds), or it may represent only a temporary phase, i.e. its development does not proceed far, and whatever be the extent of development does not proceed far, and whatever be the extent of development it becomes completely absorbed by the growing embryo (as in all non-endospermic seeds).

There are two main types of endosperm development, viz. the nuclear type and the cellular type. There is a third type also, called the helobial type, noticed in some cases.

(a) *Nuclear Type:* In this type the endosperm nucleus immediately after double fertilization divides repeatedly without corresponding wall formation. As a result a number of free nuclei appear in the embryo-sac (free nuclear phase). In a large embryo-sac several hundreds of such free nuclei may be formed, while in a small sac divisions are few. Soon a large central vacuole appears in the embryo-sac and cytoplasm together with the fee nuclei is pushed to the periphery of the sac.

(b) *Cellular Type:* In this type as the endosperm nucleus divides there is corresponding formation of cell-wall round each nucleus. The free nuclear phase, which is characteristic of the previous type, is absent free.

(c) *Helobial Type:* In this type after the first division of the endosperm nucleus a transverse wall is formed between the two newly formed nuclei, separating the embryo-sac into two unequal portions. The smaller portion lies towards the antipodal end, while the bigger portion. The smaller portion lies towards the antipodal end, while the bigger portion occupies the major area of the sac. In the latter portion, by free nuclear divisions as in the nuclear type, a number of free nuclei appear. Formation of cell-wall then proceeds inwards, finally giving rise to the endosperm. In the smaller portion the nucleus may not divide at all or it divides a few times only. If wall formation takes places here, a small mass of cells may be seen adhering to this end of the endosperm.

On the basis of seed development, there are two types of seed as

i) Endospermic seed in which a part of the endosperm persists at maturity. The endosperm, present in the mature seed, serves as food storage organ. Testa and endosperm are the two covering layers of the embryo. The amount of endosperm in mature seeds is highly species-dependent and varies from an abundant endosperm layer (*Nicotiana tabaccum*) to a single layer (*Arabidopsis thaliana*)

ii) Non-endospermic seed where the endosperm is completely consumed during embryo development the cotyledons absorb the food reserves from the endosperm. The endosperm is almost degraded in the mature seed and the embryo is enclosed by the testa. The cotyledons serve as sole food storage organs as in the case of pea (*Pisum sativum*). In terms of size seed may be as big as double coconut and as small as microscopic seeds of orchids.

In terms of structure seeds may be of two types as i) monocot seed, and ii) dicot seed.

A. Dicot seeds

Parts of Exalbuminous Seeds (gram, pea, country bean, gourd, tamarind, mango, sunflower, etc.) Seed-coats to two layers or integuments, united or free, the outer being called testa and inner tegmen, and are provided with hilum (representing the point of attachment with the stalk), micropyle (a minute opening above the hilum) and raphe (a ridge formed by the funicle or stalk in many seeds). Embryo lying within consists of an axis and two fleshy cotyledons laden with food material. The pointed end of the axis is the radicle and the feathery or leafy end the plumule. As the seed germinates the radicle gives rice to the root and the plumule to the shoot.

Parts of Albuminous Seeds (Castor, papaw, custard-apple, four o'clock plant etc.) Castor Seed. Seed coat or testa is the outer hard, blackish and mottled shell. The outgrowth formed at the micropyle is the caruncle; it absorbs moisture and helps germination. Hilum is almost hidden by the caruncle. The raphe is prominent. Perisperm is the thin white papery membrance surrounding the endosperm. This is a remnant of the nucellus and is a nutritive tissue. Endosperm is the fleshy food storage tissue, rich in oil, lying immediately within perisperm.

Embryo lies embedded in the endosperm and consists of an axis with a distinct radicle pointing outwards and an undifferentiated plumule, and two thin, flat cotyledons with distinct veins.

B. Monocot seeds

Parts of Albuminous seeds (rice, maize, wheat, onion, palms, etc.)

Each grains consists of the following parts:

1. Seed-coat is the brownish membranous layer adherent to the grain. This layer is made up of the seed-coat and the wall of the fruit fused together.
2. Endosperm forms the main bulk of the grain and is the food storage tissue of it, being laden with reserve food material, particularly starch. In a particularly starch, In a longtitudinal section of the grain it is seen to be distinctly separated from the embryo by a definite layer known as the epithelium.

3. Embryo is very small and lies in a groove at one end of the endosperm. It consist of only (a) one shield-shaped cotyledon known as the scutellum, and (b) a short axis with (i) the plumle, and (ii) the radicle protecte by the root-cap. The plumule as a whole (growing point and foliage leaves) is surrounded by a protective sheath called the plumule-sheath or coleoptile; similarly the radicle (including the root-cap) is surrounded and protected by a root-sheath called the coleorhiza. The surface layer of the scutellum lying in contact with the endosperm is the epithelium; its function is to digest and absorb food material stored in the endosperm.

In cereals (e.g. rice, wheat, maize, barley and oat), millets and other plants of the grass family the cotyledon is known as the scutellum. It supplies the growing embryo with food material absorbed from the endosperm with the help of the epithelium.

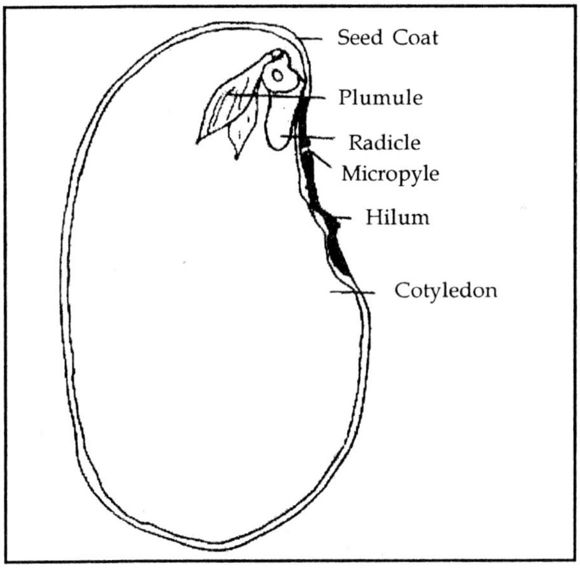

Diagramatic Sketch of Representative Dicot Seed

Monocotyledonous seeds are mostly albuminous; a few exalbuminous ones are orchids, water plantain (*Alisma*), arrowhead (*Sagittaria*), Naias, etc.

A true seed comprises a living embryo, stored food and a protective cover called seed coat. However, there are some differences between an angiospermic seed and gynnospermic seed. An angiospermic seed consists of embryo which develops from fertilized egg cell of the embryo sac, the endosperm arising from the triple fusion of two polar nucleii and one sperm cell and the seed coat developing from the integuments surrounding the ovules. Whereas the gymnospermic seed comprise endosperm or storage tissue which is haploid in nature and is modified megagametophyte. Another type of seed found is of apomictic seed which are produced asexually.

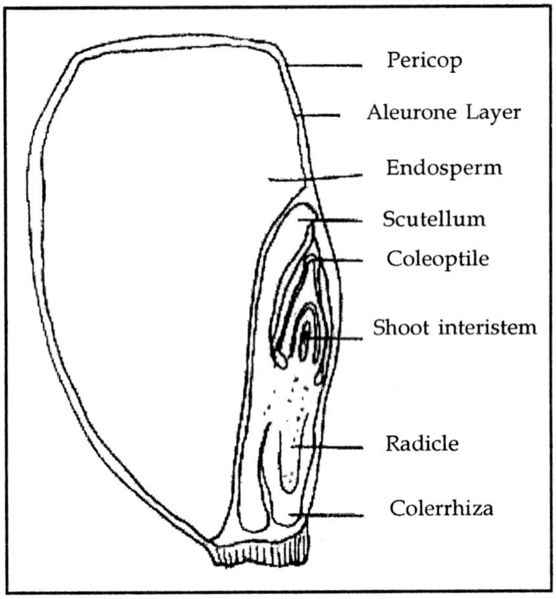

Diagramatic Sketch of Representative Monocot Seed

Embryo

Living embryo is the most important part of the seed. It consist of embryonic axis and cotyledons. The embryonic axis has embryonic root or radicle, the hypocotyle and the plumule or first leaves. In monocot seed there is only one cotyledon which is reduced and modified to form a scutellum while a dicot seed has well developed two cotyledons. The shape and size of the embryo varies from species to species. In case of orchids, the embryo is a non differentiated mass of cells.

Endosperm

A seed may be defined as endospermic seed if endosperm is present or as a non-endospermic seed if the endosperm is absent. The endosperm provides food reserves to the germinating embryo or to the seed. The endosperm is surrounded by a thin Aluerone layer which is responsible for the release of enzymes that mobilises the endospermic reserves. In general endosperm is the storage tissue but in few species cotyledons are main storage organs whereas in few perisperm, which is a diploid maternal tissue originating from nucellus, is the mani food storage tissue e.g. *Beta vulgaris, Piper nigrum.*

Seed coat

Seed coat or testa is the outer most covering of the matured seed. It shows a greater degree of variation in terms of structure, texture and colour which are under genetic control or on environmental influence. In general, all seed posses a scar like point called hilum, the attachment point to the funiculus. The micropyle, a small hole at the end of the hilum. Some times few special outgrowths are present at the hilum region as special seed adaptive structures.

Special seed structures

- *Hilum and funiculus:* Funicular scar on seed coat that marks the point at which the seed was attached via the funiculus to the ovary tissue.
- *Micropyle:* The Micropyle is a canal or hole in the coverings (seed coat) of the nucellus through which the pollen tube usually passes during fertilization. Later, when the seed matures and starts to germinate, the micropyle serves as a minute pore through which water enters. The micropylar seed end has been demonstrated to be the major entry point for water during tobacco seed imbibition and germination. During germination the tobacco testa ruptures at the micropylar end and the radicle protrudes through the micropylar endosperm.
- *Chalaza:* Non-micropylar end of the seed. The base of an ovule, bearing an embryo sac surrounded by integuments.
- *Raphe:* Ridge on seed coat formed from adnate funiculus.

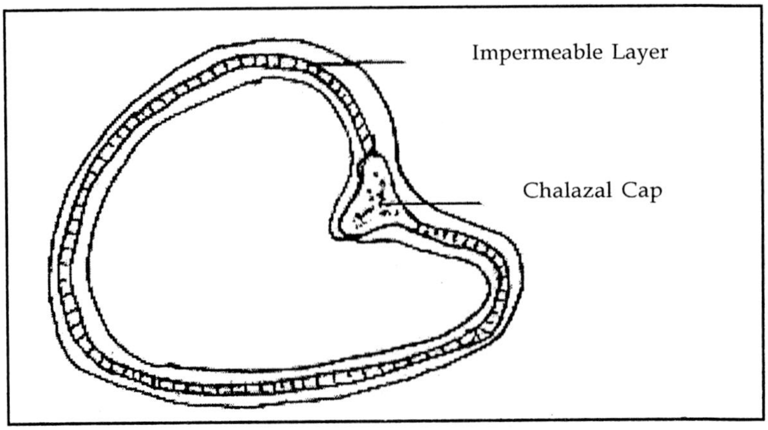

Malvaceous Seed

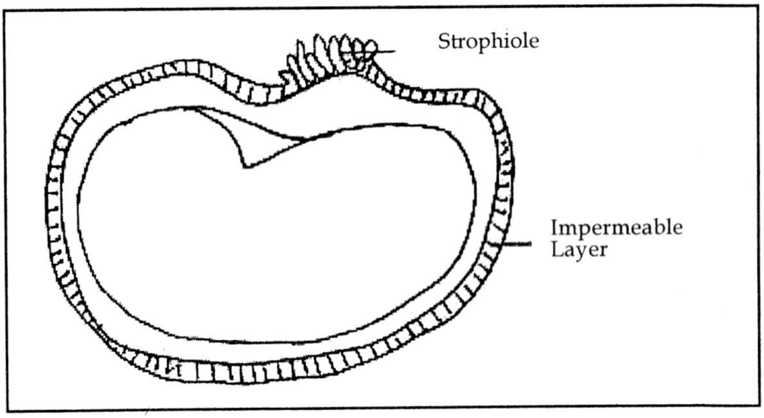

Legumenous Seed

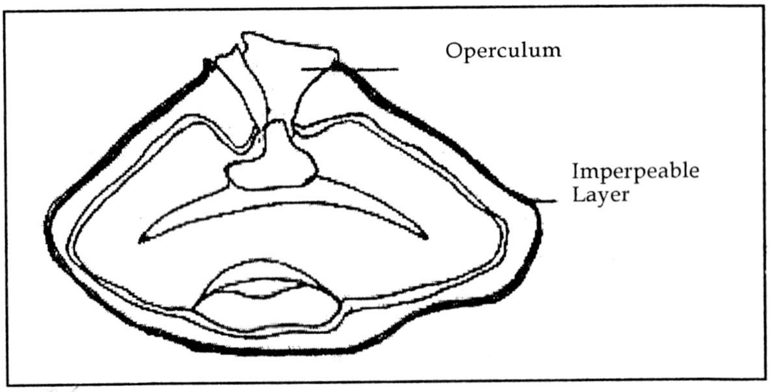

Musaceous Seed Specialized
Structures on Seed

- *Arillate:* General term for an outgrowth from the funiculus, seed coat or chalaza; or a fleshy seed coat.
- *Aril:* Outgrowth of funiculus, raphe, or integuments; or fleshy integuments or seed coat, a sarcotesta. Arils probably often aid seed dispersal, by drawing attention to the seed after the fruit has dehisced, and by providing food as an attractant reward to the disperser. The aril of the nutmeg produces the spice mace and the seed itself is the nutmeg.
- *Strophiole:* Outgrowth of the hilum region which restricts water movement into and out of some seeds. In some hard-coated legume seeds, e.g. *Melilotus alba* and *Trigonella arabica,* a plug covering a special opening - the strophiolar cleft - must be loosened or removed before water can enter, and then only through this region.
- *Operculum:* A little seed lid. It refers to a dehiscent cap of a seed or a fruit that opens during germination. e.g. *Musa* seeds.
- *Carunculate:* Seed with an excrescent outgrowth from integuments near the hilum, as in *Euphorbia.*
- *Fibrous:* Seed with stringy or cord-like seed coat, as mace in *Myristica.*
- *Funiculus:* Seed with a persistent elongate funiculus attached to seed coat, as in *Magnolia.*
- *Strophiolate:* Seed with elongate aril or strophiole in the hilum region.
- *Elaiosomes:* A specialty in the dispersal through animals is that through ants (myrmecochory). Such seeds or fruits bear attachments, the elaiosomes that contain lures and nutriments. Myrmecochory is common with plants that live at the forest soil like violets (*Viola*).
- *Caruncle:* A reduced aril, in the form of a fleshy, often waxy or oily, outgrowth near the hilum of some seeds e.g. *Ricinus, Baliospermum.* Usually it is brightly colored. It acts as an aid to dispersal. *Viola* seeds have an oily caruncle and are sought and dispersed by ants.
- *Mucilage:* A layer of polysaccharide slime produced by some seeds upon imbibition. Serves in water uptake during imbibition and germination e.g. *Ocimum* seeds.

•••

6

Fruits

The Fruit

As the ovule matures into seed, the ovary starts developing into fruit. A mature fertilized ovary is called a true fruit. If any other part of the flower is involved in the fruit development, it is called as false or pseudocarpic fruit. In some plants like banana and grapes fruits may develop without fertilization and thus are called as parthenocarpic fruits. Botanically speaking a fruit is a matured ripened ovary containing one or more ovules that mature into seed and may also include accessory floral parts. The fruits have beeen classified as simple, multiple or aggregate fruits depending upon whether they are derived from single ovary or multiple ovaries of same flower or several ovaries of a aggregate of flowers.

Simple fruit : It develops from a single flower having a single ovary developing from one or more carpels.

Aggregate fruit : It develops from a single flower having several free ovaries all of which develop into an aggregate of simple fruit.

Composite or multiple fruit : It develops from a inflorescence i.e. a group of flowers having specific arrangement gives rise to one fruit.

Types of Seeds and Fruits Found

Achene : It is a single seeded fruit developing from a superior monocarpellary ovary having marginal ovules, and its pericarp is free from the testa e.g. Ranunculus.

Cypsella : It is a single seeded dry fruit that develops from a bicarpellary, syncarpous,inferior ovary having a single basal ovule, e.g. members of compositae.

Samara : It is a one seeded dry fruit developing from mono- or polycarpellary, syncarpous superior or inferior ovary where the pericarp of matured fruit gets expanded into one or more wings,e.g. Hiptage.

Caryopsis : It is a one seeded dry fruit which develops from a tri-carpellary, syncarpous, superior ovary having single basal ovule wherein pericarp is inseparably fused with the testa,e.g. wheat.

Pod : It develops from a monocarpellary superior pistil having marginal placentation and it dehices along both the sutures e.g. members of papilionoidae.

Follicle : It also develops from a monocarpellary superior pistil but it dehices along one suture only,e.g.Calotropis procera.

Siliqua : It develops from a bicarpellary superior ovary having parietal placentation with one chamber only but it becomes bichambereddue to the development of a false septum of replum,e.g. mustard and radish.

Silicula : It is a modified siliqua which becomes short and flat having only one or two seeds e.g.Iberis.

Capsule : It develops from bi- or poly-carpellary syncarpous ovary (superior or inferior) where seeds are dispersed by rupturing of fruit in various wayse.g. through pores in Poppy.

Schizocarpic fruit : These are multi-seeded simple fruits which gets divided into one seeded compartments called the mericarps.

Lomentum : It is a large legume divided into single seeded compartments e.g. Acacia and Mimosa.

Carcerulus : It is a modified capsule divided into single seeded compartments which librate their seeds through apical pores e.g. Abutilon and Althea rosea.

Regma : It is a modified capsule which becomes divided into single seeded compartments that breaks away with explosion e.g. Castor and Geranium.

Cremocarp : It develops from a bi-carpellary inferior ovary with two compartments each having a single seed in each mericarp which remain attach to the central axis called carpophore e.g. Fennel, Coriander.

Compound Samara : It is represented by two or three, single seeded samaras which separate for dispersal e.g. Acer and Hiptage.

Fleshy or Succulent fruits : These are characterized by fleshy pericarps which are generally differentiated into epicarp, mesocarp and endocarp e.g. mango, melons and apple.

Drupe : It is developed from a multicarpellary syncarpous ovary having basal placentation and the pericarp is divided into a stony endocarp, thick mesocarp and thin epicarp. In mango, the fleshy mesocarp is edible, whereas in coconut, the mesocarp is fibrous and provides a commercial fibre called coir.

Berry : It is derived from bi-carpellary syncarpous superior ovary having a swollen axile placenta wherein epicarp is thin, mesocarp is thick and fleshy and endocarp is not distinct, members of family Solanaceae.

Pepo : It is a fleshy fruit derived from tricarpellary syncarpous inferior ovary having parietal placentation where epicarp is thin, mesocarp is thick , endocarp is not distinctand the large number of seeds are embedded in the fleshy mesocarp e.g. members of Cucurbitaceae such as melon and watermelon.

Hesperidium : It develops from multicarpellary syncarpous inferior ovary having parietal placentation where epicarp is thin, mesocarp is thickhaving oil glands and endocarp is thin and membranous. The unicellular outgrowths from the inner surface of endocarp swell up to form juicy vesicles which represent the edible part of the fruit e.g. orange and lemon.

Pome : It is a fleshy fruit derived from multicarpellary syncarpous inferior ovary wherein the fruit becomes embedded in the fleshy thalamus which represents the edible part of the fruit e.g. apple and pear.

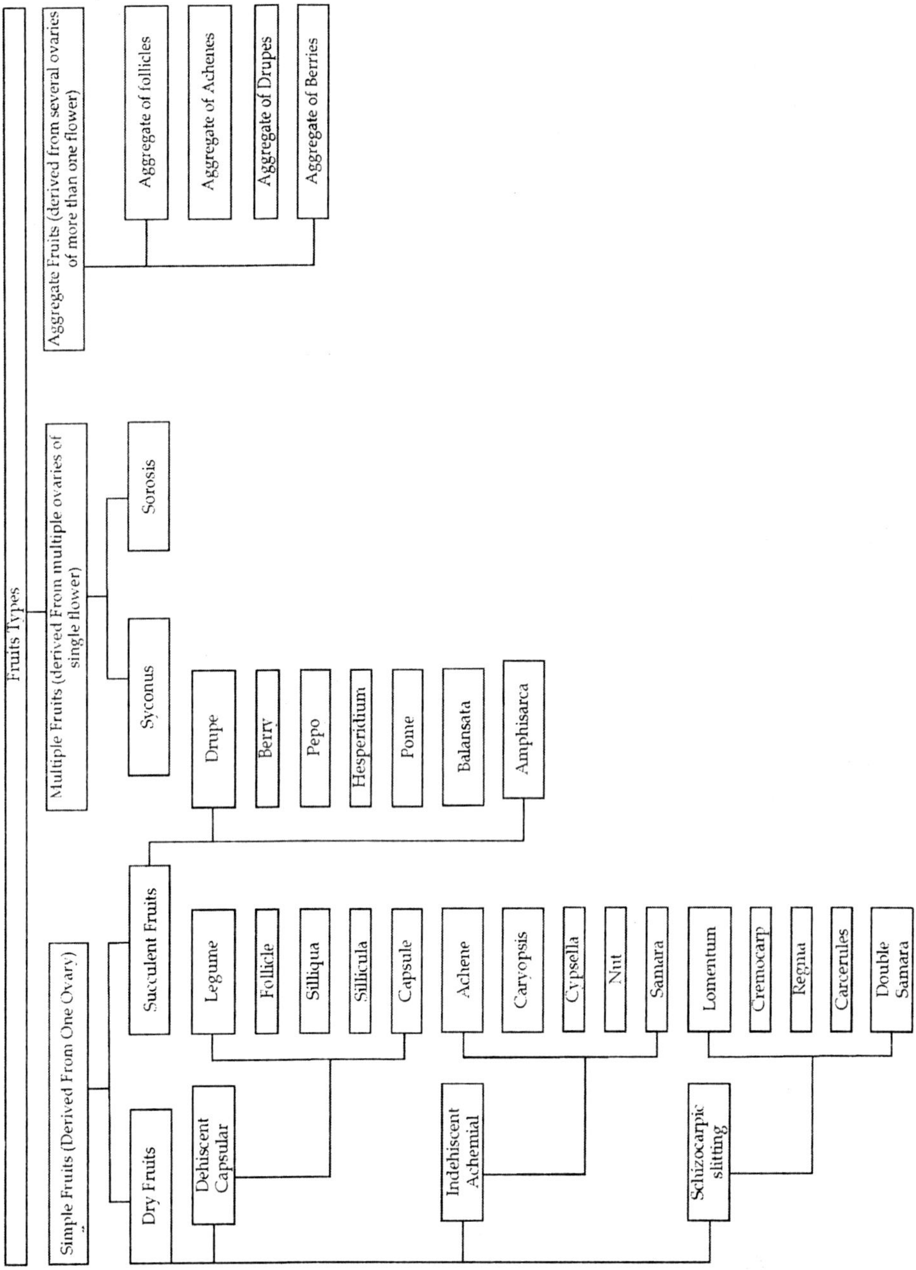
Fruits Types
Simple Fruits (Derived From One Ovary)
Multiple Fruits (derived From multiple ovaries of single flower)
Aggregate Fruits (derived from several ovaries of more than one flower)
Dry Fruits
Succulent Fruits
Syconus
Sorosis
Aggregate of follicles
Aggregate of Achenes
Aggregate of Drupes
Aggregate of Berries
Drupe
Berry
Pepo
Hesperidium
Pome
Balansata
Amphisarca
Dehiscent Capsular
Indehiscent Achenial
Schizocarpic slitting
Legume
Follicle
Silliqua
Sillicula
Capsule
Achene
Caryopsis
Cypsella
Nut
Samara
Lomentum
Cremocarp
Regma
Carcerules
Double Samara

Aggregate fruits : These are aggregates of simple fruits and are generally named as etaerious. Thus an aggregate of achenes is found in strawberry, *Ranunculus*, lotus, *Clematis* and *Naravelia*; aggregate of follicles in milk weed (*Calotropis*); aggregate of drupes in *Rubus*, aggregate of berries in Custard apple and aggregate of samaras in *Ailanthus*.

Composite or multiple fruits : Two main types of composite fruits are recognized:

Sorosis : It is derived from a catkin, spike or spadix where axis becomes fleshy and fleshy sepals also fused to form edible part of the fruit e.g. mulberry and pineapple.

Syconus : It is derived from hypanthodium inflorescence wherein the cup-shaped peduncle becomes fleshy and encloses the achenes developing from unisexual female flowers.

•••

7

Seed Drying

Three factors are of fundamental importance in controlling the longevity of seeds-water, temperature and oxygen. Seeds have been classified into two categories with regard to their response to water: orthodox or recalcitrant. Most orthodox seeds can be dried to low moisture contents (mc)<10% on wet weight basis without damage to viability and over a wide range of moisture contents there is a negative logarithmic relation between seed moisture content and longevity. Harrington's rule of thumb (1972) states that between 5 and 14% water if the seed mc is increased by 1% the life span of orthodox seeds will be reduced by 50%. The importance of drying seeds to prolong seed longevity is therefore evident.

Also it should be noted that wet seeds are more affected by extremes of temperature than dry ones. At low temperature freezing injury occurs and at high temperatures secondary dormancy is induced. Wet seeds are more liable to be attacked by insects, mites and fungi. Thus the moisture content of accessions on receipt must be determined to ascertain whether drying is necessary and if so, to determine the loss in moisture that is required. The drying process should begin as soon as possible after receipt of seeds to avoid deterioration.

To satisfy the preferred storage conditions recommended by IBPGR advisory committee on seed storage it will normally be necessary to dry accessions to 3-7% moisture content (wet basis) (depending upon species) for long-term storage except in certain cases where it has been shown that low mc causes problems for e.g. soybeans should be dried to 8% mc. On the other hand, with many species it is possible to go down to 1.5% mc without apparent damage. Nevertheless, there may be problems with severe cracking at such low mc. There are some misconceptions regarding the minimum safe mc in legume seeds. It is said that large grain legume seeds are killed below 8-9% mc but this is not strictly true. Rather than drying *per se* it is the rapid water uptake upon imbibition which is injurious; this can largely be overcome by humidification in a saturated atmosphere before imbibitions proper.

Seed banks require a drying procedure that not only minimize seed deterioration but that is also simple to operate and apply to the many different seed lots of different species which the seed bank handles.

The IBPGR recommends that for drying seeds to the recommended moisture level, a drying room maintained at about 15^{0}C and 10-15% relative humidity with good air circulation be used. This environment can be achieved by an air dehumidifier (sorption type) with refrigeration to lower the temperature and remove the heat generated by the air dehumidifier. Seeds are exposed to this environment in thin layers on open trays. Forced ventilation is used to reduce the time taken for drying. For many species (e.g. the small seeded vegetables) in open trays drying to below 6% occurs fairly rapidly over about 10 days. However, for larger seeds e.g. maize, beans etc. drying is slower and longer periods of exposure are needed. To reduce drying times for the larger seeds an alternative solution is to use a two stage drying system in which during the first stage, the drying room is maintained at a higher temperature than the one recommended, with about 40-45% RH using conventional air conditioning system (i.e. refrigeration dehumidifier). This would dry high oil content or starchy seeds to about 7-12% equilibrium mc (wb). During the second state, final drying to below 6% mc can be achieved in a self contained drier (sorption type).

The Equilibrium Moisture Content (EMC)

When seeds are held in a series of atmospheres of different relative humidities, the moisture content of the seed gradually comes into equilibrium so that the final moisture content relates to relative humidity. Equilibrium is attained either by absorption or desorption depending on the seed mc-air vapor pressure differential. Accordingly the curve is known as absorption or desorption curve and are not exact mirror images. The difference between these two curves is called hysteresis effect. A range of different RH values are obtained using a set of saturated salt or acid solution. Determining when the seeds are actually at equilibrium can be difficult if there is a large sampling error. The equilibrium moisture content (emc) values plotted against % RH at a given temperature give a sigmoidal form of curve and are termed isotherms. Three distinct regions can be demonstrated which reflect the water binding relationship of the seeds. In the low humidity range (about 0-20%) the curve is concave to the RH axis. This represents water that is tightly bound to large protein and starch molecules. In the mid range of isotherm (about 20-60% RH), the curve is usually linear, and the wateris less tightly bound. In the high humidity range the curve is convex to RH axis, rising sharply with increase in RH above 80%. This water is considered to be free water and is easily removed during drying.

Although RH is the major factor in determining emc the temperature also has an effect. Both % RH and % saturation are strongly temperature dependent. Other variables include varietal differences, seed maturity and deterioration, oil and protein content and accuracy and repeatability of the RH and moisture measuring procedures.

Factors Affecting Drying

The factors which are positively related to the rate of drying are temperature, vapor pressure, differences in oil content and air flow rate. Seed size is negatively related to rate of drying. In addition certain features of seed coat e.g. testa in legume seeds can affect the rate of drying.

Temperature not only determines the emc but also the rate of drying. Lower the temperature greater the mc of the seeds and longer the drying takes; an exponential relationship is involved. Uniformity of emc within a seed lot requires either thin layer drying or sufficient thorough ventilation to break up the microclimate. In general, seed lot should be spreads out in shallow layers rather than piled up in deep ones. Where seeds are in a thin layer the surface of seeds rapidly comes into equilibrium with the moving air.

Seeds with different oil content show different emc at a particular RH. The greater the oil content the lower the seed emc at set temperature and RH.

Rate of water loss from a seed depends very much on the structure of seed. For example, legumes where seed size is generally large and seed coat can vary in structure. In pea and beans for example, when dried rapidly, drying occurs initially from the seed coat but not from the center which remains at a higher moisture content. As drying progresses unequal drying eventually causes cracking. To avoid these problems larger seeds should be dried slowly. The proximity of the seed's moisture content to its equilibrium value also influences the rate of drying. The change in seed mc during drying slows down exponentially as the equilibrium is approached.

Seed Drying Nomogram

In the gene bank a rapid method for predicting drying times is required in order to avoid viability losses, limit seed wastage and to help select and operate drying equipment efficiently. The seed drying nomogram draws together, within a single chart, the primary factors influencing thin layer drying rate. These are size (expressed as 1000seed wt.), oil content, air velocity and drying temperature relative to the moisture content ratio; but size and drying temperature dominate the process. On the basis of moisture content ratio it is possible to predict how much moisture needs to be removed in a set time. Thus the drying nomograph chart can be used as a design tool.

Psychrometric chart

Air contains varying amounts of water vapor. The drying process is dependent on the amount of water vapor held in the air and the temperature of air/water vapor mixture. There is a limit to the maximum amount of moisture that can be held within a given volume of air. A psychrometric chart is a simple method for describing the properties of humid air i.e. air and water vapor mixture. Dry bulb temperature is shown on the horizontal axis and the air mc per kg of dry air is shown on the vertical axis. The saturation line designates the maximum moisture content in the air. The virtually vertical lines represent constant temperature. The lines for constant relative humidity show the ratio between the partial pressure of water vapor at one point and the partial pressure of water vapor in saturated air at the same temperature. Psychrometric charts can be used to illustrate viability losses under different drying conditions.

Drying equipment

Moisture can be removed from air to reduce relative humidity by using absorption or refrigeration type dehumidifiers or a combination of both.

a. Sorption type dehumidification system

Moisture can be extracted from the air by hygroscopic salts in two ways, by absorption or adsorption. Lithium chloride is an bsorbent where as silica gel is an adsorbant. In both cases materials are regenerated for reuse by heating, continuous dehumidification is usually achieved by passing the moist air through one half of a rotating drum containing the desiccant. Heated air is passed through the other half to regenerate the desiccant, and exhausted to the outside air. An alternative is to switch between two separate machines one dehumidifying while the other is regenerated.

b. Refrigeration dehumidification system

The refrigeration dehumidifier works on the principle that water vapor is condensed out of the air by cooling below the dew point temperature (i.e. the temperature at which saturation occurs). The cold saturated air is then reheated, using the latent heat removed by condensing air moisture, to provide a lower relative humidity. Refrigeration dehumidifiers are economical to operate at ambient temperatures because latent heat removed from the air moisture satisfies a significant part of the reheating load. The relative efficiency (expressed in terms of moisture removal) of typical absorption, adsorption plus refrigeration and refrigeration air pre drying application particularly, under humid tropical conditions (RH > 40% which might otherwise damage certain design of sorption based dehumidifiers. These are considered to be uneconomic for air temperatures below 15^0C or relative humid less than about 40-45%.

•••

8

Seed Purity Testing

A laboratory test measures the purity, Germination capacity, moisture percentage and other attributes of the sample regarding impurities such as weeds or dead seeds. Seed sampling for the purpose of laboratory tests refers to obtain a sample which has all the constituents of the bulk sample in same proportion. In practice, however, a seed lot is never completely uniform and, if handfuls are taken at different points, the components occur in different proportions. It is mainly because of segregation by gravity of heavy and light seeds within the bulk or within a bag or due to differences within the crop from which the seed was harvested. To obtain a sample, numerous small quantities, called **primary samples**, are taken at random from different points in the lot and mixed thoroughly to form a **composite sample**. Small quantities are then taken from composite sample and mixed to form a **submitted sample**. This process is repeated until a sample is obtained of the size required for testing, this is known as the **working sample**.

Testing for purity

In practice it is not possible to remove all these admixtures completely with the use of cleaning machines. To access the exact percentage of the different attribute, purity analysis tests are performed. In the purity test the particles are arranged in three groups components as Pure seed, seed of other species and inert matter. The normal procedure for testing purity is to draw a sub-sample of 2000-2500 seed size from the sample submitted to the laboratory and transferred to a dark smooth surface. Every seed or particle is then examined by the analyst, identified and put into one of these fractions. When the separation is complete, each fraction is weighed, and its weight expressed as a percentage of the total weight. Analytical purity is the percentage of pure seed determined in this way.

Objectives of the purity test

The object of the purity test is defined as follows:

a) To determine the composition by weights of the sample being tested and by inference the composition of the seed lot. In other words, the composition of the sample is ascertained and the results are expressed as weight percentages. When the sample has been drawn from the lot in a representative way, these results also apply to the lot.

b) To determine the identity of the various species of seeds and inert matter particles constituting the sample. It means that all seeds and other particles not only have to be separated from the good crop seed and determined by weight as prescribed in (a), but also that all seeds have to be given a scientific name.

The three components

Pure seed: Pure seed is the seed of the species that the sample purports to be, no attempt being made to distinguish between cultivars within the species. A whole seed is regarded as pure seed even if it is undersized, immature, diseased, or has germinated in store. In the case of some cereal and most grass species, the caryopsis is hidden within the chaffy floret, but the analyst has to confirm that it is present; even if its development is only partial, the floret is regarded as pure seed, if the piece is judged to be more than half the original unbroken seed. Empty florets and broken pieces less than half size are treated as inert matter.

Pure Seed Definations (From Felfoldi, 1983)

Only the structure that are described in the pure seed definition (PSD) are classifed as pure seed. All other structures among the seeds or attached to the seed have to be classifed either as other seed or as inert matter, depending on the nature of that structure.

The PSD's have been numbered for the purpose of simplicity

1. • Seed, with or without testa.

   • Piece of seed larger than one-half the original size, with or without testa.

2. • Seed, provided a portion of the testa is attached.

   • Piece of seed larger than one-half the original size, provided a portion of the testa is attached.

3. • Spikelet, with lemma and palea enclosing a caryopsis, with or without awn, plus attached sterile

   • Floret, with lemma and palea enclosing a caryopsis; with or without awn.

   • Caryopsis .

   • Piece of caryopsis larger than one-half the original size.

   N.B. 1. Separate spikelets consisting of two fertile florets.

   2. Do not separate units where the lemma of the outer sterile floret partly envelops the inner fertile florets.

   3. Remove stalk at point of attachment.

   4. Single florets containing only the ovary are classed as inert matter.

4. • Floret, with lemma and palea enclosing a caryopsis, excluding entire awn when the length of the awn is longer than the length of the floret.

   • Caryopsis.

   • Piece of caryopsis larger than one-half the original size.

5. • Spikelet with glumes, lemma and palea enclosing a caryopsis, excluding entire awn when the length of the awn is longer than the length of floret.

- Floret, with lemma and palea enclosing a caryopsis, excluding entire awn when the length of the awn is longer than the length of the floret.
- Caryopsis.
- Piece of caryopsis larger than one-half the original size.

6. • Fascicle of 1-5 spikelets * with involucre of bristles. (* Spikelet with glumes, lemma and palea enclosing a caryopsis, plus attached sterile lemma.)
- Floret, with lemma and palea enclosing a caryopsis.
- Caryopsis.
- Piece of caryopsis larger than one-half the original size.

7. • Caryopsis.

8. • Fertile (Sessile) spikelet ,* (with or without awn) with accached pedicel (or sterile spikelet) and rachis segment (stalk), provided that the stalk is not longer than the spiklet.

(when the stalk is longer than the spikelet, the entire stalk is to be removed.)

(* The fertile spikelet consists of hardened glumes enclosing two florets, therefore not visible, one consisting of a transparent sterile lemma, the other containing the caryopsis within a tissue-like lemma and palea, with or without awn.

- Caryopsis.
- Piece of caryopsis larger than one-half the original size.

9. • Cluster, or piece of cluster, including attached stalk* pieces not protruding from the cluster, more than the width of the cluster, retained on a 200mm x 300mm rectangular sieve with square ended slots 1.5m x 20mm when shaken for one minute, with or without seed.

(* protruding stalk pieces that are longer than the width of the cluster, are to be entirely removed.)

10. • Cluster, or pieces of cluster, including attached stalk* piece not protruding from the cluster more than the width of the cluster, unless it is obvious that no seed is present.

(*protruding stalk pieces that are longer than the width of the cluster, are to be entirely removed.)

- Seed, with the pericarp/testa partially or entirely removed.
- Piece of seed larger than one-half the orifinal size with the pericarp/testa partially or entirely removed.

11. • Achene, with or without pappus, unless it is obvious size, unless it is obvious that no seed is present.

- Piece of achene larger than one-half the original size, unless it is obvious that no seed is present.
- Seed, with the pericarp/testa partially or entirely removed.
- Piece of seed larger than one-half the original size with the paricarp/testa partially or entirely removed.

12. • Schizocarp/mericarp, with or without pedical (of any length or frequency), unless it is obvious that no seeds are present.

- Piece of mericarp larger than one-half the original size, unless it is obvious that no seed is present.
- Seed, with the pericarp, partially or entirely removed.
- Piece of seed larger than one-half the original size, with the pericarp partially or entirely removed.

13 • Seed, with or without testa.*

- Piece of seed larger than one-half the original size, with or without testa.

14 • Seed, with or without testa, with or without strophiole/ caruncle.

- Piece of seed larger than one-half the original size, with or without testa.

Other Seed: Into this fraction are put seeds of any species other than the pure seed species, and each seed has to be identified. This fraction includes both crop seeds and weed seeds, and the percentage of any species is reported if it is greater than 0.1. Empty florets (and broken seeds less than half-size) are treated as inert matter, as in the case of the pure seed species.

Inert matter: This fraction takes everything in the sample that is neither "pure seed" nor "other seed". It includes seed material such as empty grass florets and broken seed pieces half-size or less; other plant material such as chaff, awns, leaves, and pieces of stem; mineral matter such as soil, sand, and stones. For convenience it also includes fungus bodies such as ergot, and insect galls, even though (strictly speaking) these are not "inert".

Equipments needed for purity analysis

1. A magnifying glass or a binocular microscope may be needed for the small-seeded species or when one wants to determine presence of species that are hard to distinguish from the crop seed.
2. Weighing balances intended for weighing samples, sub-samples, fractions and components must meet certain requirements regarding precision. An analytical balance enabling 0.1 mg to be read accurately (Capacity 160 - 200g) and a precision balance with about 1 kg capacity, allowing 10 mg to read accurately. with a tare mechanism are desired.. Weighing
3. *Working table:* All balance should be placed on a weighing table. Such a table should consist of a strong slab (8cm thick) resting on anti-vibration cushioning and supported by concrete or brick pillars. Weighing tables can be purchased from the balance manufactures, or be made locally. A concrete floor or a brick wall is needed at the places where the weighing table will be located.
4. A set of small hand-sieves with different perforations can be very useful for dividing the working sample into two or more portions that are more homogeneous. This facilitates the test in two ways: it will be more accurate and will take less time to analyse.
5. Miscellaneous equipments like spoons, spatulas, forceps, scalpel needless, shallow trays, funnels, watch glasses, etc. are always needed for a normal purity analysis.
6. Reference seed collection should be maintained properly in labelled glass test tubes and stored in the drawers. A small quick reference collection for each analyst of crop seeds and weed seeds in specimen tubes that are arranged in holes in a wooden block will save a lot of time and effort. It is made by the analysis themselves with seeds that they found in the samples.

Working procedure of the purity test

(a) Working Samples

The purity analysis shall be made on a working sample taken from the submitted samples.

(b) Separation

- The sub sample (purity working sample) is spread on the working table;
- Each particle is judged individually, the criteria used being: external appearance (shape, size, colour, gloss, surface texture) And/or appearance in transmitted light;
- All other seeds and inert matter particles present are removed leaving the pure seed, the separations resulting in the three components mentioned above;
- Seed enclosed in fruits other than those indicated in pure seed shall be separated and the detached empty fruit classed as inert matter;
- After separation, each component part and any species of seed or kind of other matter of which a percentage is to be reported, shall be weighed in grams to the minimum number of decimal places and recorded on the purity.
- Components may be retained for future reference although the pure seed will be sent to the germination unit, where 400 seeds of it will be used for the germination test.

Data Analysis

The purity test conducted on one whole sample or two halves of one whole sample or two whole samples and the results are expressed as follows

1. *Weight Gain or Loss During Analysis:* Add together the weight of all the component fractions from the working sample, This sum must be compared with the original weight as a check against gain or loss. If there is a discrepancy of more than 5% of the initial weight, a retest must be made. The result of the retest is then reported.

2. *Calculation of component percentage:* The percentage by weight of each of the component parts shall be given to one decimal place . Percentage must be based on the sum of the weight of the components, not on the original weight of the working samples.
3. The percentage of seed of any particular species other than the pure seed, or of any particular kind of inert matter need not be calculated except as required for reporting result.
4. *Rounding Procedure:* Add together the percentage of all fractions. Fractions that are to be reported as a "trace" are excluded from this calculation; the other fractions shall then together total 100.0% If the sum does not equal 100.0% (either 99.9 or 100.1) then add or subtract 0.1% from the largest value (normally the pure seed fraction). (Note: If a correction of more than 0.1% is necessary, check for a calculation error.)
5. Test for variation between two halves working samples or two whole samples is done using the averages of the various components and it should be within the limits of tolerances.

Reporting results

The result of a purity analysis shall be given to one decimal place and the percentage of all component must total 100. Component of less than 0.05% shall be reported as "Trace". The percentage of pure seed, other seed and inert matter must be reported in the spaces provided on the Analysis Certificate. If the result for a component is nill, this must be shown as '-0.0' in the appropriate space.

The latin name of the species of pure seed must be reported on the Analysis Certificate. The kind of inert matter and the Latin name of each species of other seed must be reported.

•••

9

Seed Moisture Estimation

The survival of the seed in dry storage depends upon its moisture content more than on any other factor. The sensitivity of seed longevity to moisture content also provides a way to prolong the life of the seed by reducing the moisture content. Since seed moisture and its management is very important activity for a genebank, its estimation should also be done very critically. Moisture testing procedures can be classified as primary or secondary methods. In primary or direct methods of moisture estimation, the water is removed and the amount is determined quantitatively. These are destructive in nature but provide greater accuracy. Secondary or indirect methods are those in which some chemical or physical property of the seed is related to moisture content. These measurements are calibrated against the true moisture content determined by a primary method. These are less accurate but are much faster and are more practical for field use.

Methods

There are four different type of methods available for moisture estimation as follows

a) Lyophilization (freeze-Drying)

b) Chemical desiccation (vacuum ; without heating)

c) Karl Fischer's Titration Method

d) Air-Oven Drying Method

Karl Fischer Method

The Karl Fischer method for determining moisture content is based on the titration of the methane-water mixture with the Karl Fischer reagent (a solution containing methanol, iodine, sulphur dioxide and pyridine). Any water present reduces the iodine to colorless iodine. The titration end point is the first appearance of brown iodine colour. The Karl Fischer reagent is specific for water with no interfering substances in the seed extracts that react with the solution. This is therefore, probably the most basic and accurate method available. Two types of automatic Karl Fischer titrators are available commercially. Both employ electrometric end point detection. These are capable of measuring moisture contents from 0 to 1gH2Og-FW and can complete the titration in <10min. Methods of performing karl Fischer moisture determinations on seeds with automatic Karl Fischer titrators are described by Benjamin and Grabe (1988). The utility of this method in genebank is that this can be used to calibrate the moisture meters or other methods used for rapid me estimation.

Air-Oven Method

Air-Oven Method is the most acceptable method among all and included in the International Rules for Seed Testing Association (ISTA), which is further modified by ISTA on the basis of the seed characteristics as follows

a) Low Constant Temperature Oven Method. (Recommended for oily seeds).

b) High Constant Temperature Oven Method. (Recommended for the other type of seed).

Materials required

1. Containers of non-corrosive metal/glass moisture bottles
2. An electrical oven with adequate Ventilation and a thermostatic
3. Analytical balance that weighs accurately in grams upto three decimals places (precision of I mg)

4. A grinding mill constructed of material that cannot absorb moisture (wood in not suitable).
5. A desiccator with a suitable desiccant such as silica gel.
6. A brush and steel brush to clean the mill after grinding.
7. Metal trays with holes
8. Tongs and a cloth.
9. A small spoon to transfer the seeds into the container.

Procedure

a) Pre-drying

- If seed moisture is more than 17% (10% in case of soybean and 13% for rice) pre- drying is obligatory (ISTA rules).

b) Grinding

- Grinding is obligatory to bold seeds
- Low constant temperature Oven method : Generally followed for oily seeds (modified ISTA)
- Weigh the moisture bottle along with the lid and number them (same number on lid and bottle), (w1)
- Add approximately 4-5g (0.5g in case of less seeds) of prepared sample and weigh accurately along with lid (w2)
- Remove the lid and keep the moisture bottles in oven for 17$\pm$ hour at 103°C
- Remove the bottles from the oven; replace their lids and place in a desiccator having silica gel to cool for 30 minutes at room temperature.
- Remove the bottles from desiccator and weigh them one by one (W3).

c) High Constant Temperature Oven Method

It is generally followed foe non-oily seeds. The procedure is same as described for low constant temperature oven method except that the oven is maintained at a temperature of 130^0 -132^0 C for 1-4 hours.

Calculation of results

The moisture content (MC) is calculated to two decimal place.

$$\% \text{ moisture content} = \frac{\text{Weight of fresh seeds} - \text{weight of dry seeds}}{\text{Weight of fresh seeds}} \times 100$$

$$\text{Mathematically} = \text{MC }\% = \frac{W2 - W3}{W2 - W1} \times 100$$

Where

W1 = weight of container with cover

W2 = weight of container with cover and seeds before drying.

W3 = weight of container with cover and seeds after drying.

For samples, which have been pre- dried, use the following formula

$$\text{Final }\% \text{ moisture content} = \frac{(S1 + S2) - \{(S1 \times S2)\}}{100}$$

Where

S1 = Percent moisture content from 1st stage of drying.

S2 = Percent moisture content from 2nd stage of drying.

Precautions

- All the accessions entering the drying room should be clearly labeled to prevent mixing. Seeds should be dried in thin cloth bags or any such pervious material. Large sized accession should be subdivided into small bags for drying rather then being left in a single large bag. Seed bags should not be stacked too closely but should be spread on open racks in the drying room or cabinet.
- The use of 15^{0}C and 15% RH is compromise between cost, rate of drying and loss in viability. Therefore, these conditions should be maintained to avoid deterioration of seeds during the drying period.
- Movement of seeds from the drying room should be done carefully to avoid condensation of moisture on surface.

- Although insects do not generally breed at low humidity prevalent in the drying rooms, adequate care should be taken to screen incoming samples for infestation to check spread to other accessions and to maintain general sanitation inside the drying room.
- It is advisable that stand- by dehumidifiers and air conditioners are available in the event of equipment failure.
- Moisture arriving in the drying room through cracks/leaks can considerably raise the drying load and thus the room should be pressure tested for leaks.
- According to ISTA rules the difference between moisture determination of two replicates should be no greater than 0.2%./ The test should be repeated id the difference is greater.
- The total weight of seed used for a moisture content determination should be sufficient to make the test accurate and yet not be wasteful of seeds. The ISTA rules for seed testing recommend that two replicated of 4g of seed, however, this weight of seeds is not appropriate for use in genebanks because large number of small seeds is needed, unnecessarily depleting the accession sampe size. Therefore, it is suggested that approximately 0.5g of seeds per replicate can be used for moisture content determination in genebank.

•••

10

Seed Quality Enhancement

Seed Quality Enhancement

Seeds are the delivery systems for agricultural biotechnology. High quality seed leads to excellent seedling performance in the field. Seed quality is a complex trait that is determined by interactions between multiple genetic factors and environmental conditions. Modern approches to improve seed quality therefore combine classical genetics, plant molecular biology and a variety of seed technologies. These "seed biotechnologies" enhance physiological quality, vigor and synchronity to establish a crop in the field under diverse environmentalconditions.

Seed technologies for quality enhancements include priming, pelleting, coating, artificial seeds. The main focus of the treatments is on embryo growth and on the different seed covering layers (e.g. testa, endosperm, pericarp), which are determinants of seed quality and exhibit the biodiversity of seed structures. Seed germination is controlled by environmental factors (light, temperature, water) and on plant hormones as endogenous regulators (gibberellins, abscisic acid, ethylene, auxin, cytokinins, brassinosteroids). The utilization of plant hormones and inhibitors of their biosynthesis and action in seed treatment technologies affects seed germination and seedling emergence. The genes, enzymes, signaling components and down-stream targets of plant hormones provide molecular marker for seed quality and seedling performance.

Seed priming is the most important physiological seed enhancement method. It is an hydration treatment that allows controlled imbibition and induction of the pre-germination metabolism ("activation"), but radicle emergence is prevented. The hydration treatment is stopped before dessication tolerance is lost. An important problem is to stop the priming process in the right moment. Priming solutions can be supplemented with plant hormones or beneficial microorganisms. The seeds can be dried back for storage, distribution and planting. Germination speed and synchrony of primed seeds are enhanced and can be interpreted in the way that priming increases seed vigor and a wider temperature range for germination, release of dormancy and faster emergence of uniform seedlings is achieved. This leads to better crop stands and higher yields. A practical drawback of primed seeds is often a decrease in storability and the need for cool storage temperatures.Several types of seed priming are commonly used:

Osmopriming or osmoconditioningwhere the seeds are incubated in well aerated solutions with a low water potential, and afterwards washes and dried. The low water potential of the solutions can be achieved by adding osmotica like mannitol, polyethyleneglycol (PEG) or salts like KCl.

Hydropriming or drum priming is achieved by continuous or successive addition of a limited amount of water to the seeds. A drum is used for this purpose and the water can also be applied by humid air. 'On-farm steeping' is the cheep and useful technique that is practized by incubating seeds (mainly the cereals and legumes) for a limited time in warm water.

Matrixpriming or matriconditioning is the incubation of seeds in a solid, insoluble matrix (vermiculite, diatomaceous earth, cross-linked highly water-absorbent polymers) with a limited amount of water to facilitate slow imbibition of water in the seed.

Pre-germinated seeds is only possible with a few species. In contrast to normal priming, seeds are allowed to perform radicle protrusion. This is followed by sorting for specific stages, a treatment that reinduces dessication tolerance, and drying. The use of pregerminated seeds causes rapid and uniform seedling development.

Film-coating methods allow the chemicals to be applied in a synthetic polymer that is sprayed onto the seeds and provide a solid, thin coat covering them. The advantage of the polymers is that they adhere tightly to the seed and prevent loss of active materials like fungicides, nutrients, colorants or plant hormones. Some novel applications of film coating are used to modify imbibition and germination. They can confer temperature-sensitive water permeability to seeds or affect gaseous exchange. By this they control the timing of seed germination and seedling emergence. Certain temperature-dependent water-resistant polymers can delay imbibition till the environmental conditions become suitable for germination.

Seed pelleting adds thicker artificial coverings to seeds, which can be used to cover irregular seed shapes and add chemicals to the pellet matrix, e.g. of sugar beet or vegetable seeds. The pellet matrix consists of filling materials and glue. Loam, starch, tyllose (cellulose derivative) or polyacrylate/polyacrylamide polymers are commercially used. A film coat can be added onto the pelleting layer as shown in the figure above. It is also used to increase the size of very small horticultural seeds. This provides improved planting features, e.g. singulate planting, the use of planting machines, or precise placement and visibility in/on the soil.

•••

11

Seed Germination

Germination involves the activation of the quiescent, resting embryo. Consequently for germination to occur, the seed must absorb water which is sufficient to initiate germination. The first sign of embryo growth is the prutusion of some part of the embryo, mostly the radicle through the seed coat. Thus radicle emergence is the clear objective and criterion for germination. However a seed capable of achieving radicle emergence may not produce a seedling capable of independent growth (independent that is of stored food reserves within the seed) such that a healthy plant can be produced. To a seed analyst, germination in a laboratory test is the emergence and development from the seed embryo of those essential structures which for the species being tested, indicates the ability to develop into a normal plant under the favourable conditions in the soil.

Importance of the Germination Test

Generally if the seed to be tested is clean, viable and non-dormant, the result from such a test performed under optimal laboratory conditioned will correlate positively with the field emergence. In order to bring uniformity in testing procedures, International Seed Testing Association (ISTA) formulated a set of rules. However, for genebank purpose IBPGR Advisory Committee on seed has also formulated a set of ruleswhich are basically ISTA rules with slight modifications. Any genetic resources laboratory should have both stes of rules to guide them for the germination testing procedures.

Types of Germination

Germination has been classified as hypogeal germination or Epigeal germination on the basis of the position of the cotyledonsduring germination in relation to the soil surface.In Hypogeal germination, the cotyledons do not emerge above the soil surface. This type of germination is characteristic of peas and grasses. In Epigeal germination the cotyledons or the comparable storaghe structures emerge above the soil surface. There is rapid elongation of the hypocotyl, which pushes the cotyledons and the plumule through the soil. This type of germination is charateristic of beans and pine seeds.

In the Gene bank, viability of a seed lot is a measure of how many seeds in the lot are alive and could develop into normal, healthy plants when provided with suitable conditions. Viability of a seed lot/ accession is usually measured in order to asses its suitability for a particular purpose, most commonly to produce a normal healthy crop but sometimes also for other purpose as in malting of barley to work out the most suitable steeping procedures. Successful stand establishment requires high quality, genetically pure seeds that produce rapid and uniform seedlings. An array of quality tests are in use ranging from genetic purity to seed vigour analyses. Proper storage conditions maintain a high germination percentage. The most common and reliable method of testing seed viability is the germination test. Germination is defined as the "emergence and development from the seed embryo of those essential structures which, for the kind of seed tested indicate its ability to develop into normal plant under favorable conditions in the soil" (Anonymous 1985).

The conditions for laboratory germination, therefore, must not only be precise enough to initiate seed growth, but must also favor seedling development, in a limited period of time, to a stage in which all essential structures could be evaluated, to enable the separation of normal seedlings from abnormal ones. Germination can therefore, be expressed as the percentage of pure seeds of a given type of seed that produces normal seedlings under optimal conditions.

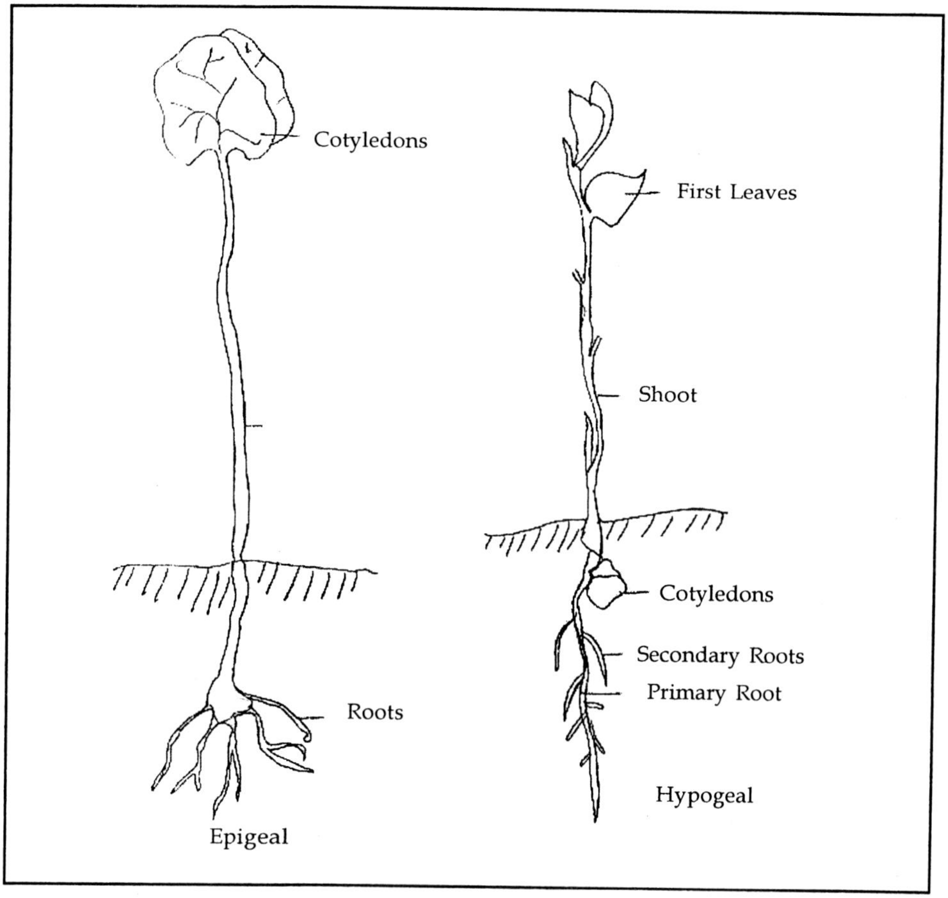

Types of Germination

General Principles

The germination test in the laboratory should always be done on pure seed fraction. A random sample of 400 pure seeds should be taken and put for germination; in replicates of 100, 50 or 25 seeds. The seeds should be uniformly spaced on a moist substratum. Since germplasm collected is very valuable and sometimes only limited quantity of seed is available under such circumstances IBPGR Advisory Committee recommends that for the initial germination test for the species where a reasonable germination technique is available, a minimum two replicates using 200 seeds (100 seed per replicate) is acceptable, providing that the germination is above 90%. If not a further 200 seeds should be tested as before and the overall result for seed viability taken as mean of the two tests.

The replicates are then placed under optimal germin-ation conditions, usually including a treatment to break dormancy, if needed. The first count is made when the majority of seedlings have reached the developmental stage at which proper evaluation is possible. Since the seedlings being tested are entirely dependent for growth on the nutrients stored within the seed, they must be evaluated before they exhaust the nutrients and begin to rot.

The normal seedlings are removed and counted. Rotten seeds and decayed seedlings are also removed to prevent further contamination. At the time of final recording, number of fresh ungerminated and hard seeds is also counted. At the end of the test if the result of the replicates falls within the tolerance range, the average of normal seedlings of the replicates represents the percentage germination of that seed lot/accession.

Types of Substrata

Depending on the laboratory germination facility and type of the seed, its size and requirement various kinds of substrata could be employed for germination testing. Generally substrata used in germination test include paper, sand, soil and agar. However, the accuracy and reproducibility of the germination results are very much dependent on the quality of the substrata used for germination testing. The germination substrata should therefore, meet the following basic requirements.

a. It should be non-toxic

b. It should be free of molds and other micro-organisms & their spores.

c. It should be able to hold enough moisture for the germination of seeds & their development to a suitable stage.

d. It should provide adequate aeration to the germinating seeds.

e. It should be easy to handle and use finally.

f. If should be less expensive

g. As far as possible the pH of the substratum should be between 6-7.5.

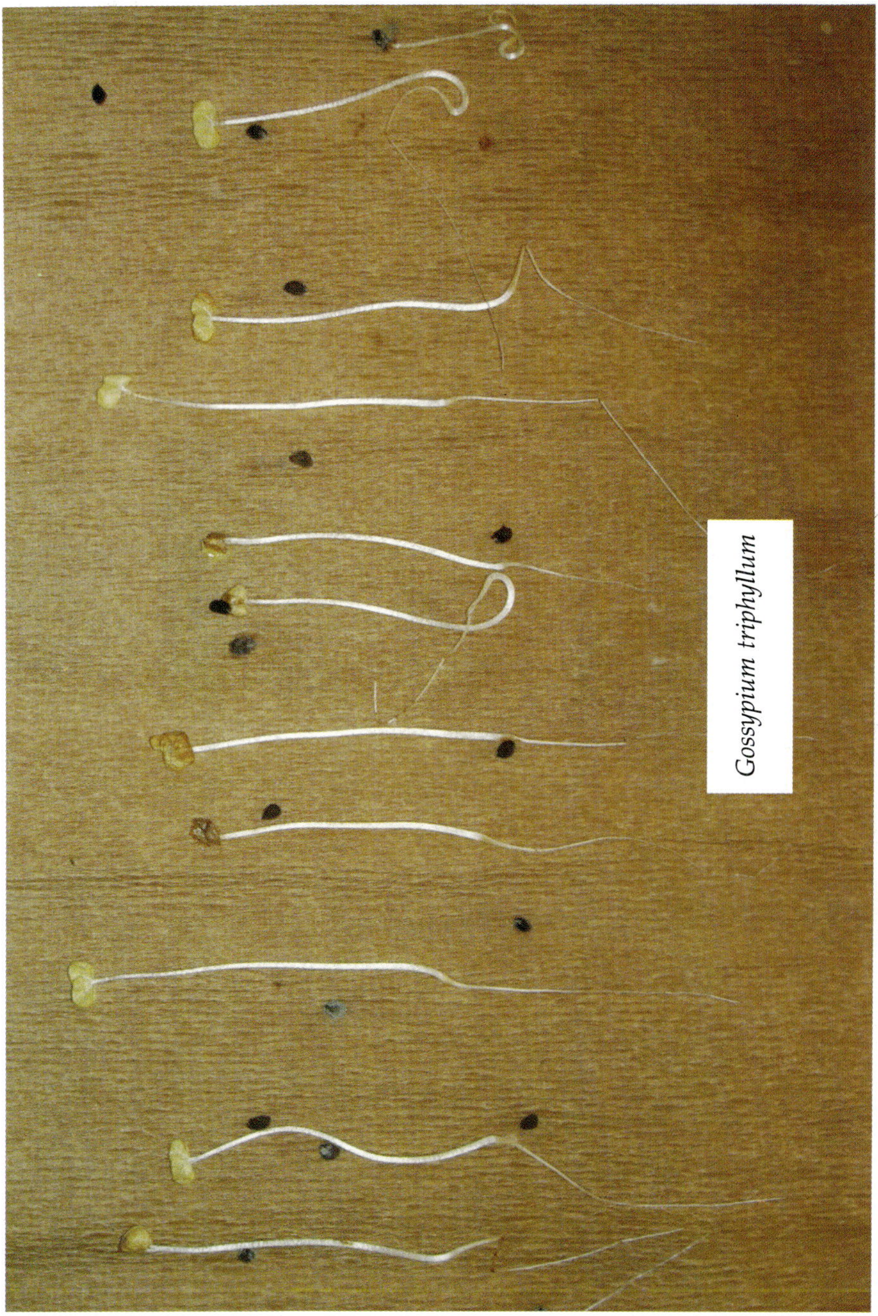

Seed Germination Using Towel Paper Method (BP)

Seed Germination Using Patriplates (TP) Method

(A) Paper

The paper used for germination should be of good quality and free from toxic dyes and other substances. (ISTA rules 1985). Before germination testing the paper should be soaked in running water for 2 hours so as to moisten it evenly and to remove water soluble toxic substance, if present. For most seeds, the paper should be sufficiently but not excessively moist. (The moist paper on being pressed between the thumb and fore finger should not form a film of water.) Paper may be used for germination in one of the several ways listed below.

1. T.P. (Top of Paper)

The method is generally practiced in case of small seeds. Seeds in this case are germinated on top of two or more layers of paper in either a Petri dish or some other container with a air tight lid to prevent the loss of moisture.

Steps

Paper of known quality such as Whatman filter paper should be used.

1. The paper should be cut in the form of discs according to the size of the Petri dish.
2. Soak the paper in water and put 2-3 layers in the Petri dish having airtight lids. Hold the Petri dish in slanting position to drain off excess water.
3. Record the lot no./ accession number and date on piece of paper and stick it to the lid from inside. Space the counted seeds on the moist paper and cover the lid.
4. Transfer the test to a germinator at the desired temperature.
5. Observe and record the germination data after prescribed interval.

2) B.P.(Between Paper)

The method is generally used for medium and large seeds such as those of cereals, legumes, vegetables etc. Seeds are germinated between two or more layers of germination paper. This method is also known as the rolled towel method.

Steps

1. The towel paper is soaked in running water. When uniformly and sufficiently moist, the paper is removed. It is then pressed to remove the extra water and placed in a plastic tray.
2. Two layers of wet towel paper are placed over a sheet of wax/ butter paper. (The glossy side of butter paper should be inside. On the top corner of right side write the lot/accession no, date of test and name of the crop).
3. Arrange the seeds on top of the moist paper towel either with hand or with the help of a counting board.
4. Cover the seeds by placing another layer of moist paper towel either with hand or with the help of a counting board.
5. Turn up two inches of the bottom edge to form a lip (this prevents the seeds from falling out.)
6. The towels are gently rolled ensuring no extra pressure is placed on the seeds and tied with a rubber band.
7. The prepared rolled towels are then placed in a basket in a slanting position and transferred in a germinator maintained at desired temperature.

P.P. (Pleated papers)

The test is generally carried on small seeded crops which required good amount of moisture.

Steps

1. A germination paper strip is folded into 10-50 pleats longitudinally.
2. Both the ends of the paper are stapled.
3. Seeds are placed in the pleats and sufficient water is added.
4. The strips are either kept in plastic boxes with air tight lids or placed directly in a 'wet' type seed germinator.

(B) Sand

Sand is normally used as substrate for lager seeds such as castor, groundnut, beans etc. Depending on their size, the seeds can either be planted on a layer of moist sand and covered with 10-20 mm of loose sand, or planted on the top of the most sand and pressed into it. The sand should be properly graded, sterilized and free from impurities and toxic chemicals. The sand particles should be able to pass through a sieve of 0.8 mm diameter holes and be retained by a sieve of 0.05mm diameter holes.

The amount of water to be added to the sand will depend upon its characteristics and the size of the seed to be tested. The optimum amount should be determined for the main kind of seeds, so that a measured quantity could always be added in routine testing. Generally for cereals except maize, sand is moistened to 60 per cent of its water holding capacity.

The approximate quantity of water to be added to the sand can be calculated by the formula given below.

$$\text{ml of water to be added to every 100g of sand} = \frac{118.3\text{ml sand} \times (20.2-8)}{\text{wt of 118.3ml of sand (in gms)}}$$

Soil

Soil or artificial compost is commonly used instead of sand to test sample that produce seedlings with phyto-toxic symptoms when germinated in sand or paper. The soil or artificial compost is generally more difficult to standardize and is therefore, liable to cause greater variation between results. But this substratum must be used to confirm the evaluation of seedlings in doubtful cases and for testing samples which produce seedlings with phyto toxic symptoms when germinated on paper. Water should be added until the consistency of soil is such that the ball formed by squeezing it in the palm of the hand is easily broken when pressed between two fingers.

Germination environment

After placing the seeds on the prescribed substrata, the replicates should be transferred to the controlled temperature and relative humidity conditions maintained in cabinets or walk-in germinators for the prescribed period which varies according to the species.

(a) Temperature

Temperature is one of the most critical factors in laboratory germination testing. In the rules for seed testing, two kinds of temperature conditions are provided. A single numerical indicate the constant temperature e.g. 30°C and a numerical separated by a hyphen (-) indicate alternating temperatures e.g. 20-30 °C. Usually, when alternating temperatures are used, the lower temperature is maintained for 16 hours a day and the higher temperature for eight hours. If alternating temperatures cannot be conveniently attained, the tests must be maintained at lower temperature. The temperature should be kept as uniform as possible with variation not exceeding +-1°C in each 24 hours period.

(b) Moisture and aeration

The moisture requirement of seeds will vary according to its kind. Large seeded species generally required more water than small seeded species. It is essential that the substratum is kept sufficiently moist throughout the germination period. Care need to be taken that the substratum should not be too moist. The excessive moisture will restrict the aeration and may cause the rotting of seedling. Except the situations where a high moisture level is recommended (e.g. paddy and jute), the substratum should not be so wet as to form a film around the seeds.

The water used for moistening the substratum should be reasonably free from acidic, alkaline and other organic and inorganic impurities. The pH of the water should be in the range of 6.0-7.5. Under condition where such water is not available it is advisable to use distilled or deionised water. Uptake of water is an essential step to initiate the germination. The imbibition of water is determined by the chemical composition of the seed, permeability of the seed coat. Sometimes seed absorbs 2 to 3 times water when compared to its dry weight. It is completed in three phases as

Phase 1: Seed just mechanically absorbs water enough for metabolic activities to start.

Phase 2: Active phase in which nil absorption but the metabolic activities are at maximum rate.

Phase 3: There is little increase in the uptake of the water just to combat the ongoing activities.

Special measures for aeration are not usually necessary in case of top of paper (TP) tests. However, in roll towel test (BP) care should be taken that the rolls should be loose enough to allow the presence of sufficient air around the seeds. In case of soil and sand the care should be taken to avoid crust formation or compaction of top most layer of the substratum.

(c) Light

Seeds of different crops have varied light requirement. Some seeds germinate only in the dark, others require light, and others are not affected by light. Seeds which require light for germination must be illuminated with cool fluorescent light or with day light. For at least 8 hours in every 24 hours cycle, with alternating temperatures, light is provided during high temperature period. The light intensity should be between 750-1250 lux.

Far red light negatively affects germination of many seeds, while red light often promotes germination. Cool white fluorescent lamps are ideal because they emit relatively less of far red light and more of red light and do not cause appreciable heating to affect the germination temperature.

Morphological and Physiological changes during germination

Germination is the most active phase in the plant survival. It involves numerous processes of which major ones are as follows

- Imbibition of water
- Hydration of the subcellular organelles
- Changes in the subcellular organisation of the embryo & endosperm or cotyledons
- Changes in the phytochrome activity
- Activation of the enzymes
- Denovo synthesis of the enzymes
- Digestion of the food reserves
- Translocation of the soluble organic molecules to embryo
- Synthesis of proteins& other constitutents
- Increase in the O2 uptake and respiratory activities
- Cell enlargements

- Cell division
- Synthesis or the Activation of the growth substances
- Cellular differentiation
- Redistribution of the metabolites with in the embryo
- Changes in the O2 & CO2 levels

Evaluation of Germination Test

In evaluating the germination test, the seedlings are categorized either as normal or abnormal seedlings. For categorizing the seedlings as normal or abnormal they need to be evaluated at the right developmental stage where all essential structures can be thoroughly inspected and before seedlings run out of stored nutrition in the endosperm or cotyledons etc. whenever a sample produces results which can't be readily evaluated, another test should be made in good quality sand under favorable conditions.

Seeds put for germination are classified into one of the following categories as per ISTA rules (Anon., 1985).

Normal Seedling

Those seedlings which show capacity for continued development into a normal plant when grown in good quality soil with optimum supply of water, light and temperature. Normal seedling can be completely intact or have slight defect or secondary infection which does not interfere with normal growth and development. (detailed classification can be had from ISTA Handbook for Seedling Evaluation (Bekendam & Grabe 1979).

Generally speaking normal seedlings should have:

A) A well developed root system including a primary root, except in certain species of family Graminea where seminal roots are produced.

B) A well developed and intact hypocotyl and or epicotyl without any damage to conducting tissue and in dicots a normal healthy plumule.

C) In Gramineae, a well developed primary leaf within a coleoptile.

D) One intact cotyledon for monocots and two cotyledons for dicots. Seedling with following slight defects, provided they show vigourous and balanced development of other essential structures.

E) Seedling of *Zea mays*, species of Cucurbitaceae and Malvaceae with damaged primary roots but with sufficient lateral and adventitious roots to support seedling growth and development.

F) Seedlings with superficial damage or decay to essential structure which otherwise does not affect normal growth of seedlings.

G) Seedlings which are seriously decayed (fungi, bacteria) but only when it is clearly evident that seed was healthy and did not carry any infection on/within it and that all essential structures were present.

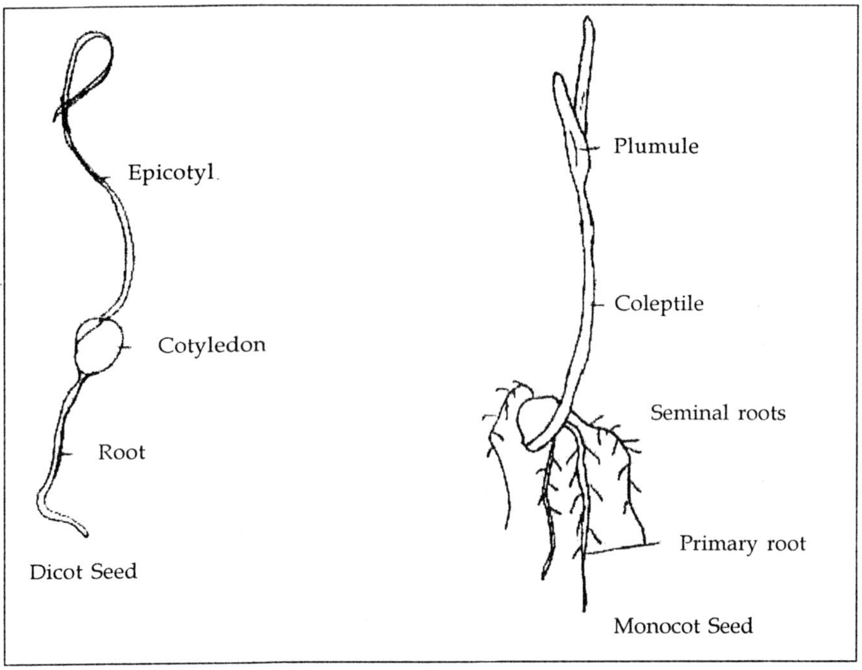

Normal Seedlings

Abnormal Seedlings

Those seedlings which do not show capacity for continued development into normal plants when grown under favorable conditions of light temperature and water. Seedling with following defects can be classified as abnormal seedlings.

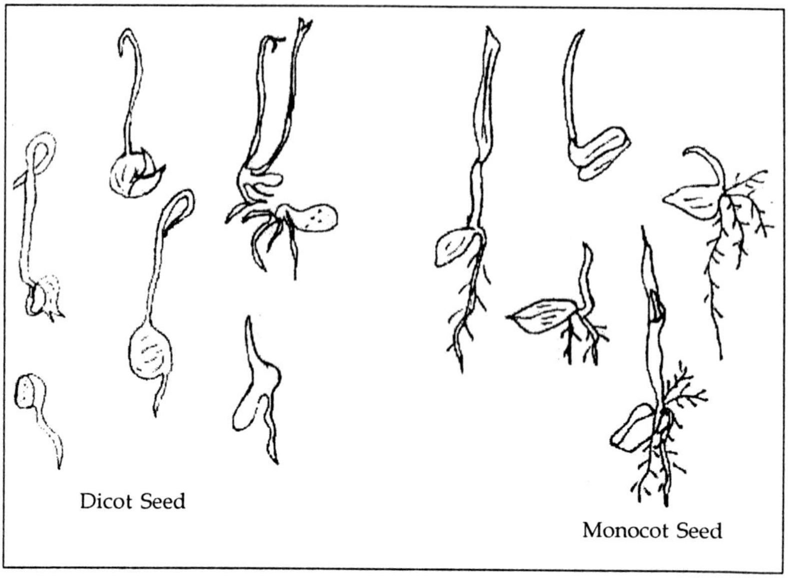

Abnormal Seedlings

1) Damaged seedlings

A) Seedling with badly damaged/bruised cotyledons

B) Seedlings with splits cracks and lesions of essential structures

C) Seedlings with no primary roots in those species where primary root is essential.

2) Deformed seedlings

A) Seedlings with weak or unbalanced development such as spirally twisted or stunted plumules, hypocotoyls and epicotyl.

B) Swollen shoots

C) Stunted roots

D) Split plumules,

E) Empty coleoptile

F) Watery and glassy seedlings.

3) Decayed Seedlings

A) Decay in essential structures resulting from seed borne infection.

Hard Seeds

Seeds which remain hard and dormant at the end of the prescribed test period because they have not absorbed water due to an impermeable seed coat, are classified as hard seeds.

Fresh ungerminated seeds

Seeds, other than hard seeds which remain firm and apparently viable, even after appropriate treatment of breaking dormancy in accordance with ISTA rules are classified as fresh ungerminated seeds.

Dead seeds

Seeds which at the end of the test period are neither hard nor fresh and have not produced seedling are classified as dead seeds.

Every seedling must be evaluated in accordance with prescribed rules, for evaluation, the stage of development and essential structures must be sufficient to permit detection of the abnormal seedlings. In case of unsatisfactory results arising due to toxicity of substratum, recording error or due to secondary infection etc a retest may be conducted.

Special pre-printed notebook or cards are useful for recoding the results of germination analyses. After the prescribed period or on the last counting day, evaluation is completed. Hard and fresh ungerminated seeds are also recorded along with normal and abnormal seedlings. The number of seed and seedling in different categories must equal the number of seeds originally planted.

The average of normal seedlings represents the percentage germination.

Essential equipment required for germination in laboratory

Counting devices

These are required for maintaining specific conditions of temperature, relative humidity and light. These are generally of two types.

a) Cabinet germinators

b) Walk in germinators

Other requirements

The other equipments required for routine germination testing include refrigerators, scarifier, hot water bath, incubator, germination boxes, Petri dishes, forceps, plastic trays, oven for sterilization of glassware, sand/soil germination paper, beaker, measuring cylinders, scales etc. Certain chemicals like Gibbrelic acid, potassium nitrate. Thiourea, Sulphuric acid, Tetrazolium chloride, distilled water etc. are also required for specific purposes.

•••

12

Seed Dormancy

It is a common observation that seeds of many plant species do not germinate when freshly harvested even under favorable conditions. Otherwise viable and mature seeds when fail to germinate and resume growth immediately after in an environment normally favorable for the germination of the concern species is known as seed dormancy. One can define dormancy in many ways as

- Seed dormancy is nature's way of setting a time clock that allows seeds to start germination when conditions are normally favorable for germination and survival of the seedlings.
- Dormancy is a naturally occurring phenomenon of many plants to maximize the chances that seed will germinate at an appropriate time.
- Dormancy is the physical or physiological condition of a viable seed that prevents germination even in the presence of otherwise favorable germination conditions.
- Dormancy is a seed characteristic, the degree of which defines what conditions should be met to make the seed germinated.

The problems of seed dormancy is more significant in gene banks because

- It interferes with the results of germination tests designed to assess the viability of accessions.
- It can reduce the amount of seeds which emerge during regeneration and multiplication and consequently lead to selection for less dormant types and genetic erosion of genetically heterogeneous accessions.

In gene banks, attempts are made to store seeds as soon as possible after harvest which is also when dormancy is more pronounced. Low temperature drying of seeds in gene banks not only minimizes loss in seed quality but also minimizes loss of dormancy. The sub-zero temperatures used in long term seed storage also maintain the dormant state of seeds.

Types of Dormancy

To many people seed dormancy simply means that a seed has not germinated. But why it does not, has two probable reasons. One that it does not receive favourable conditions for germination to start and secondly it must be due to some property of the seed or dispersal unit that prevents it from germination. **Nikolaeva** 1969 had classified dormancy as **endogenous** type and **exogenous** dormancy. In endogenous dormancy, some characteristic of the embryo prevents germination whereas in exogenous dormancy, some characteristic of the structures, including endosperm (sometimes perisperm), seed coat or fruit walls, covering the embryo prevents germination. These blocks are to be removed before a seed germinates. The challenge of germination ecologists is to define the environmental conditions required to bring about changes in seed that result in release from dormancy and to correlate these factors that promote dormancy break in nature.

Endogenous dormancy

It can be further classified as three types

a) Physiological dormancy

This type of dormancy is caused by a physiological inhibiting mechanism of the embryo that prevents radicle emergence. However, structures that cover the embryo, including endosperm, seed coats, and indehiscent fruit walls, may play a role in preventing germination. This type of dormancy can be divided into three main categories as non-deep physiological, intermediate and deep physiological dormancy. Non deep physiological dormancy is very common in seeds of most weeds. Freshly matured seeds of many vegetables and garden flowers cannot germinate at any temperature or they germinate only over a narrow range of temperatures. This type of dormancy is broken by relatively short periods of cold stratification ranging from 5 days in *Triticum* sp. to 60 to 90 days in *Impatiens biflora*. Also when the seeds stored dry at room temperatures, they come out of dormancy. How ever, the time required for seeds to after ripen usually is much longer than that required for dormancy loss during cold stratification. Seeds of *Digitaria ischaemum* require 8 weeks of cold stratification at 3° C or one year of dry storage at room temperature to become non- dormant. This type of dormancy can also be broken by chemicals, including potassium nitrate, thiourea, kinetin, ethylene and gibberellins. A light requirement for germination is another manifestation of this type of dormancy. Most of the seeds with Physiological Dormancy are permeable to water, but there are few exceptions where a combined form of physical and physiological dormancy is encountered and germination does not occur until both type of dormancy have been broken.

b) Morphological dormancy

Sometimes in few species at the time of seed dispersal, the embryonic structures like radicle and cotyledons can be distinguished, but are not fully developed. Thus embryo growth is required before germination occurs. While in seeds of other species, embryo is just a mass of cells at the time of dispersal i.e. it is undifferentiated. In both type of seeds germination does not takes place because of morphological characteristics of the embryo, hence is termed as morphological dormancy. It occurs in seeds with rudimentary and linear embryos. Underdeveloped embryos occur in seeds of a number of families that are primarily are tropical in distribution like Annonaceae, Arecaceae, Lactoridaceae, Myristicaceae and Winteraceae. Under tropical conditions underdeveloped embryos grow slowing after the seeds are dispersed, and the seed are eventually germinate. In temperate regions, morphological dormancy is reported in two families only Apiaceae and Ranunculaceae.Seeds of the cultivated de Caen type of *Anemone coronaria* have morphological dormancy, but those of wild type have both morphological and physiological dormancy. Undifferentiated embryos occur in families with micro- seeds like Orchidaceae, Pyrolaceae. In these plants either the radicular or plumular end of the embryo mass elongates and eventually emerges from seed. The tissue that emerges from the seed enlarges to form a haustorium or protocorm, depending on the species.

c) Morphophysiological dormancy

This type of dormancy occurs in seeds with rudimentary or linear embryos, and as the name indicates it is a combination of morphological and physiological dormancy i.e. the underdeveloped embryos have physiological dormancy. MPD is found in a number of plant families, including the Apiaceae, Araliaceae, Araceae, Aquifoliaceae, Aristilochiaceae, Berberidaceae, Fumariaceae, Illiciaceae, Lardizabalanaceae, Liliaceae, Magnoliaceae, Papavaraceae, Ranunculaceae and Schisandraceae (Grushvitzky, 1967).

Exogenous dormancy

It can be further classified as three types

a) Physical dormancy

In this type of dormancy, the primary reason for the lack of germination is the impermeability of seed (or fruit) coat to water. Physical dormancy is present in about 15 families of angiosperm. These families possess a palisade layer of lignified cells in the seed coat (Vazquez-Yanes and Perez-Garcia,1976; Corner,1976), a characteristic frequently found in seeds with physical dormancy. The number of these palisade layers may vary in different species and taxa e.g. impermeable seed coats in the family Fabaceae have only one palisade layer while in Convolvulaceae two palisade layers are present. There are present some specialized sclereid cells with thick lignified walls that mainly cause the impermeability of the seed coat. The most common type of sclereid cells are macrosclereid or malphigian cell.They are impregnated with water repellent substances including cutin,lignin,quinones, pectic-insoluble materials, suberin and wax. Impermeable seed coats developed in *Pisum elatius* during seed dehydration. As the dehydration process begins, catechol oxidase increases eight fold in the seed coats, but it decreases as seed reach their final dry weight. Before seed can germinate, however, an opening or passage through the palisade or other impermeable layer should be formed, whereby the water enters the seed. The type of opening varies with the plant families. In addition to the development of impermeable layers in the seed coats, all natural openings including the micropyle, hilum and chalazal area also become impermeable to water.

Plant families with physical dormancy

Name of the family	**Seed Type**	**Specialized Structures**
Anacardiaceae	Bent	Operculum
Bombacaceae	Spatulate	Micropyle
Cannaceae	Linear	Lid on raphe
Cistaceae	Bent	Chalazal Plug
Convolvulaceae	Folded	Plug near micropyle
Cucurbitaceae	Spathulate	Micropyle
Caesalpinoideae	Investing	Hilum
Mimosoideae	Investing	Strophiole
Papilionoideae	Bent	Strophiole
Gerniaceae	Folded	Hilum
Malvaceae	Folded	Chalazal plug
Musaceae	Capitate	Micropylar lid like structure
Nelumbonaceae	Broad	Protuberances

b) Chemical dormancy

According to Nikolaeva,1969, chemically dormant seeds do not germinate due to presence of inhibitors in the pericarp. This definition of chemical dormancy has been broadened to include compounds that are either produced or translocated to the seed, where they block embryo growth. Germination inhibitors have been found in the embryo, endosperm and seed coats of the seeds and in structures that sometimes are dispersed along with the seeds of some species . A complication in the study of chemical dormancy is that most seeds from which germination inhibitors have been isolated also exhibit the presence of physiological dormancy. Seeds of *Hyoscyamus muticus* contain one water soluble inhibitor but after-ripening resulted in increased germination percentage.

c) Mechanical dormancy

According to Nikolaeva (1969), mechanical dormancy is due to the presence of hard ,woody fruit wall. The woody structure usually is endocarp, but sometimes the mesocarp also is woody (Hill, 1933). Stony endocarps are found in (some or all members) many families, including Anacardiaceae, Apocynaceae, Arecaceae, Burseraceae, Cornaceae, Elaeagnaceae, Eleocarpaceae, Juglandaceae, Lecythidaceae, Meliaceae, Nyssaceae, Oleaceae, Pandaceae, Rhamnaceae, Rosaceae, Sapotaceae, Tiliaceae and Zygophyllaceae.

Secondary Dormancy

In many plant species non dormant seeds quite often become dormant due to exposure to one or more unfavorable conditions of storage, such as temperature, moisture or light. This is called secondary dormancy. In nature secondary dormancy is generally exhibited by many weed species. Seeds of these plants remain viable for many years in nature due to this property only. A good example of this is found in *Xanthium* where exposure to oxygen stress condition is responsible for the secondary dormancy induction. In case of *Isabgol* exposures to low temperature result in secondary dormancy.

Breaking of Dormancy

Dormancy is considered to be an adaptive response of the seed to adverse environmental conditions. In nature dormancy terminates as the environmental conditions become favorable for germination. The impermeable seed coat becomes softened due to the rupturing facilitated by many micro0organisms present in the soil. Sometimes seed coat become permeable due to exposure to diurnal changes in the temperature in soil. Fire and smoke, thus produced are also reported to break the dormancy of many forest tree species seeds. In addition to these natural ways of dormancy breaking, several treatments are also devised to terminate this dormancy in the laboratory.

Techniques to break dormancy

As we proceed with the germination test and if a greater proportion of fresh ungerminated seeds or hard seeds are observed, it is because of dormancy which needs special attention when it comes to conserve

distributive strategy which provide a delay mechanism which enables seed dispersal from parent before germination. It is nature's way of setting a time clock that allows seeds to start germination when conditions are normally favorable for germination and survival of the seedlings. It is a naturally occurring phenomenon of many plants to maximize the chances that seed will germinate at an appropriate time.

The survival of natural populations depends mainly on their ability to exploit the favorable conditions and avoid the unfavorable environmental conditions to which they are constantly exposed in their natural habitats. Therefore plants have developed highly effective adaptations to their environments through natural selection. The state of dormancy equips them to escape the deteriorating effects of the natural conditions, thereby enhancing their survival rates. During the period of dormancy all the physiological activities of growth and development are ceased. Other metabolic activities like respiration, photosynthesis etc. is in the suspended state. Therefore dormancy is a vital phenomenon for the survival of a species.

Seed dormancy - a greater problem in Gene Banks

While conserving the seed in gene banks, the phenomenon of dormancy needs a special attention as

- It seriously interferes with the results of germination test to assess the viability, and
- It reduces the amount of precious seed which emerges when seeds are sown out for regeneration, multiplication or other purposes.
- In order to maximize seed quality and therefore seed longevity attempts should be made to store seeds as soon as possible after harvest, which coincidentally, is when they are at their most dormant stage.
- In order to maximize seed quality, seed drying is done at a lower temperature which also minimize the loss of dormancy.
- The sub-zero temperature not only prolong viability but also the dormancy.
- Gene Bank mainly conserve the wild and primitive cultivars which generally show more of dormancy.

Is there exists a Silver Platter?

The answer is simply NO. but practically speaking, it may be possible to offer an "Aluminum Platter". The aluminum platter is an approach for determining the type of dormancy a particular species exhibits, by examining the type of habitat in which it occurred.

The Aluminum Platter

Habitat of origin	Dormancy treatment
Temperate regions by cold winters	Stratification
Short growing seasons (alpine, boreal forests)	Warm + cold stratification
Undestroyed species and small seeds	Light
Dry climate (grassland, desert)	Leaching, scarification
Warm high humidity	Scarification

It is important to note that many of the factors which are said to promote the germination of dormant seeds may only work under certain conditions. The interaction between the effects of various dormancy breaking agents on seed germination can either be positive or negative. To determine these effects, require testing of different sub samples of the accession under all possible combinations of these dormancy breaking agents. Various modifications with regard to time, duration and concentration etc. in the above methods can be made to initiate germination in various crop species. Sometimes a combination of some of above treatments can prove more effective than using only one method.

The ultimate desirable aim of the test for a given species is to be able to provide within one test environment or procedure those conditions which enable all viable seeds of that species, whether dormant or non dormant, to be able to germinate. The International Seed Testing Association (ISTA) and Association of Seed Analyst of North America (AOSA) rules summaries the available information on these aspects and are generally suitable for application in Gene banks. However, even where a prescription is well tried, gene banks will still need to be alert to the possibility that some accession may be discovered for which prescription does not appear to work. The Algorithm approach developed for genebanks is that instead of a single prescription, an attempt has been made to develop the shortest series of tests to define a suitable test environment. The decision as to which test to try next depends on the results obtained in the previous test(s).

•••

13

The Quick Viability Test

Historical Perspective

The tetrazolium test developed in the 1940s is perhaps the most significant advance in seed testing technology of this country. In the beginning Lakon developed the topographical selenite method which stressed the reaction of individual seed organs and resulted in considerable accuracy in estimating the seed viability. However this method presented considerable hazard to analysts since this chemical is highly toxic. In 1941, Kuhn and Jerchel discovered the tetrazolium salts were reduced from colourless soluble forms to insoluble colored formazans in living tissue. This discovery lead to the abandoning of the toxic selenite by Lakon in favour of the nontoxic tetrazolium salts which in turn resulted in the development of the rapid viability testing with these salts in several seeds. His first work was on corn seeds and small gains. He found that when tetrazolium slats came in contact with living tissue in seed embroyos, the chemical was reduced to an insoluble pigment and stained the tissue. Non living tissue did not effect the reduction of the salts. Of the several tetrazolium slats tested, Lakon found 2, 3, 5 - triphenyl tetrazolium chloride to be most suitable. The test was introduced into the United States after the World War II.

Mechanism of Tetrazolium reaction

Tetrazolium salt is an oxidation-reduction indicator and the development of a non-diffusible red color in tissue is the result of reduction of the chemical by enzymatic action. One or more of the dehydrogenase systems appear to be involved in the reaction. The reaction is as follows

$$\text{2, 3, 5-triphenly tetrazolium chloride} \xrightarrow{+2e+2H} \text{Formazan}$$

The dehydrogenase enzymes are involved in the respiratory activity of biological systems. During the respiratory process intermediates are produced which serve as substrata for the enzymes. Hydrogen ions are transferred to tetrazolium which acts as hydrogen acceptor. Tetrazolium is then reduced to an insoluble, coloured formazan. Since the reaction occurs within the cells and the pigment is non-diffusible, there is a rather sharp delineation between respiring (viable) tissue and non-respiring (non-viable) tissue. The former takes on a characteristic red colour while the latter retains its natural colour.

Seed and seedling structure

Intelligent and accurate application of the tetrazolium test demands at least a general knowledge of seed and seedling structure. Such knowledge is essential for proper preparation of the seeds for the test and for accurate interpretation of test results. Interpretation in primarily based on the distribution for live and dead tissue among various organs of the embryo. Unless the significance and function of each organ in known, the staining reactions are meaningless.

To the maked eye as in soybeans or in discernible as in cucurbits and clovers. The hypocotlyl-radicle axis is variously shaped : rootlike in radish and legume seed or flat and triangular as in cucurbits.

Proper development of the primary root is of great importance in dicots. If the primary root is weak or does not develop, adventitious or secondary roots may develop; however, overall plant development is adversely affected.

Obtaining the Sample

The seeds used for the standard germination test are taken from the pure seed component of a purity analysis. Since the principle objective of the tetrazolium test is to provide an estimate of germination percentage, similar procedures must be followed. The seeds used for the tetrazolium test should, therefore, also be taken indiscriminately from the pure seed fraction. The use of counting boards for large seeds or pneumatic devices for small seeds will eliminate much of bias in taking the sample.

It is very important that selection of the appropriate number of seeds for the tetrazolium test be the first step. A serious error can be introduced in the test when a different procedure is used. The appropriate number of seed for the test with a few extra should be placed in the beaker to soak. After soaking all the seeds are removed and used for test.

For most purposes, 2 x 100 seed are sufficient for the tetrazolium test. In some cases 2 x 50 or 1 x 100 seed will provide the necessary information. When only a rough approximation of viability as good, medium or poor is needed, 1 x 500 seed will suffice. Once the seeds are selected they are reedy for preconditioning or preparation for the tetrazolium test.

Preconditioning the seeds

Interpretation of staining patterns is facilitated when seeds are softened in water before sectioning or piercing. During this period, activation of enzymatic processes takes place and the tissues becomes the fragile.

Seeds are conditioned by placing them on top or between moist blotters or paper towels over night or by placing them in a beaker of water for 3-6 hours at a warm temperature (35^{0}C) depending on the kind of seeds being tested. The overnight method generally is more compatible with seed testing laboratory routine.

Seeds of small seeded legumes and some other kind don't require conditioning but may be placed directly in the tetrazolium solution.

Seeds that are heavily treated with red-coloured fungicide should be washed before conditioning to prevent the red dye fro obscuring the tetrazolium staining patterns.

Preparation of the Seed

There are several general methods for preparing seeds for the tetrazolium test. Depending on the kind of seed one of the following techniques could be employed to this purpose.

a) *Direct staining:* some seeds such as brassicas, soybean clover and vetch will stain directly through the seed coat. The air dry seeds are placed directly in the tetrazolium solution. The seed coats however, must be removed before interpretation.

b) *Removal of the seed coat:* In some species it is necessary to remove the seed coat and other membranes to permit absorption of tetrazolium. This is usually done by soaking the seed coat for considerable duration and carefully removing the seed coat with fingers for forceps and placing them in the tetrazolium solution e.g. cotton seed and bean seed.

c) *Bisecting:* Large monocotyledonous seeds such as corn are sectioned longitudinally and medially through the embryo. One half of the seed is discarded and the other half reserved for the test. Generally, it is necessary to soften the seeds by soaking them for a period of time in water so that they will not fracture or crumble when sectioned.

d) *Puncturing or severing distal end:* Seeds which are too small to bisect properly are punctured into the endosperm above the embryo with a sharp needle or the entire distal tip (non-embryonic end of the seed is severed. Preconditioning by soaking in necessary before puncturing or cutting to prevent fracturing of the seed. The seeds are sometimes preconditioned on moist filter paper in Petri plates and punctured or cut under magnification.

Staining

Seeds are stained in a small glass or plastic container such as Petri dish or beaker. Sufficient solution of desired concentration should be used to cover the seeds and to allow for the absorption of tetrazolium solution.

Several concentrations of tetrazolium solution may be used with comparable results. Conveniently, stock solutions of 1.0% and 0.1% may be maintained. The 1.0% solution is used for seeds that are not bisected through the embryo, while the 0.1% solution is used for seeds in which the embryo is bisected. The pH of the solution should be between 6 and 8 for best staining to occur. If the pH of the water is not the netural range, the tetrazolium salt should be dissolved in a phosphate buffer solution of pH 7. The solution can be dept for several months at room temperature if stored in dark or amber coloured bottle to prevent deterioration from light.

Seeds are allowed to remain in the tetrazolium solution for 2-16 hrs. The staining time depends on the kind of seed, concentration of the tetrazolium solution and temperature. Care should be taken not to over-stain seeds, as this causes degeneration, making interpretations difficult and uncertain.

After the complete staining, the tetrazolium solution should be drained out and the stained. Since tetrazolium is photoreducible, the seeds should be incubated in dark during the staining process. Seeds should be washed thoroughly with water several times in order to remove the adhering traces of the solution and to avoid staining artifacts.

Different kinds of seeds stain at rates characteristics for the kind Variation in stain intensity is greatest among the seeds that are stained directly through the seed coat and particularly among those kinds of seeds exhibiting hardseededness. Seeds with broken seed coats stain very rapidly, those with readily permeable but intact seed coats stain more slowly. Uniformity of staining can be increased by soaking the seeds for several hours in water before placing in tetrazolium solution.

Temperatures between 20 and 45°C have no effect on accuracy of tetrazolium tests, but staining proceeds faster at the higher temperatures. Tests may be performed satisfactorily at 25 °C temperatures in dark.

Evaluation

Upon penetration into living cells, the tetrazolium choloride is reduced by dehydrogenase enzymes present in the living tissue to formazan which a reddish, water-insoluble compound. The reaction occurs within living cells which are releasing hydrogen in the respiration processes. Viable tissues produce a normal red colour. Weak living tissues produce an abnormal colour. Such tissues have lost some of the initial resistance to the penetration of tetrazolium. Respiration is accelerated and formazan is produced rapidly. During early stages of deterioration, these tissues become darker red (bruised tissue) more rapidly than sound, healthy tissues. Dead tissues do not strain, remaining usually white (aged tissue) because lack of respiration prevents formazan production by embryo tissues. Depending on size, all seeds are examined under a microscope. Larger seeds such as pea may be examined without a microscope.

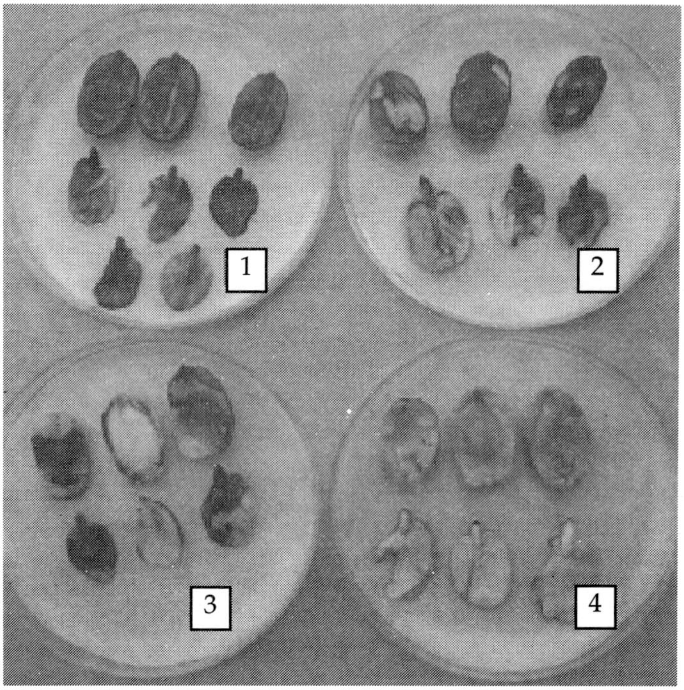

TTC Staining Pattern of *Jatropha Curcas*

(1) Seeds completely and uniformly stained
(2) Seed showing minor unstained areas on cotyledon opposite to deeply stained radicle tip and parts of cotyledon
(3) Seeds showing necrotic tissue towards the radicle tip and in the centre of the cotyledon
(4) Seeds completely unstained

Accurate interpretation of the tetrazolium chloride test is dependent upon.

- Knowledge of seed and seedling structure and seed germination
- Understanding of individual embryo structures and their interdependence in the development of respective seedlings structure.

- Understanding of the mechanism of the test and its limitations.
- Combining interpretation of staining patterns with other visible aspects of seed quality.
- Experience with making comparative germination test.

The TZ test is basically sound and widely used among many accredited laboratories in Canada and the United States as well as Europe. Accuracy of results depends largely upon the training, experience, education and background and other general qualifications of the analysist. The main function of the tetrazolium test is to distinguish between viable and non-viable seeds.

Some advantages of the TZ test are:

- A rapid evaluation of viability
- The detection of seed weaknesses before they become evident in germination tests.
- Timely guidance in quality control programs.

Some limitations of the TZ test are:

- Requires specialized training and experience
- Test is usually more laborious and tedious to perform than a germination test.
- Tests results do not reflect fungi infection
- Tests does not reflect dormancy.

Reporting of results

All tetrazolium chloride test results are recorded as a percentage (same as germination results). The tetratzolium chloride test results are usually reported in the remarks section of the Certificate of Seed Analysis issued by the seed laboratory. The statement should read:

Tetrazolium chloride test results: __________ % vaiable seeds
(test may also include non-viable%)

hard seed seeds are to be reported (if applicable) as a percentage and included in the total % viable seed. Regardless of how reported. TZ test results for viability provide an estimate of the maximum percentage seeds that have the capability (potential) to produce normal seedlings.

Accuracy of results

Properly conducted tetrazolium and germination test results are generally in close agreement, within the range of normal sampling variation. Differences of 3-5% may be due entirely to unavoidable sampling of seeded crops than with small-seeded crops and with uniform seed lots than with non-uniform lots.

•••

14

Seedling Vigour

Germination is defined as the emergence and development form the seed embryo of those essential structures, which for the kind of seed in question, indicate its ability to develop into a normal plant under favorable condition. It is many a times observed that seed lots of apparently equal quality as indicated by germination percentage produce largely different responses in field emergence. Variations such as these arise due to differences in vigour of seed in the seed lots. Clearly a germination test alone is not enough to assess the seed quality. Therefore, determining "the degree of aliveness" (vigour) of a seed a just as important as determining whether the seed alive.

Isley (1957) defined vigour as "the sum total of all seed attributes which favour stand establishment under unfavorable field conditions. However, the Association of Official Seed Analysts (Anon., 1975) Defined seed vigour as the "sum total of all those properties in seed which, upon planting, result in rapid and uniform production of healthy seedlings under a wide range of environment including both favorable and stress condition". Another definition is that of ISTA: "Seed vigour is the sum total of those properties of the seed which determine the potential level of performance and activity of a non-dormant seed or seed lot during germination and seedling emergence".

A vigour test cannot replace a germination test, but supplement it with more information about seed quality.

Various Seed Vigor Tests

TYPE OF TEST			
Physical seed size physical soundness	Performance first count speed of germination seedling growth rate seedling dry weight	Stress cold test cool germination test paper piercing test accelerated ageing test Brick gravel test RQ membrane integrity	Biochemical Glutamic acid dexarboxylase activity (GADA test) Tetrazolium test respiration and repiratory quotient (RQ)

Selected test for vigour

(1) *First count:* This test can be conducted along with the standard germination test. The number of normal seedlings counted / removed when the first count of the germination test is done is an indication of the seed lots quality ; the higher the percentage of normal seedlings removed during the first count, the better the seed quality.

If the first count is to be used to compare different seed lots over several months, all first counts must be done at the same time interval after planting.

(2) *Speed of germination:* if a more detailed test is desired, speed of germination may be recorded. This test can also be incorporated in to the standard germination test. In this case, germination is considered to have occurred when the radicle has appeared and therefore counts need to be made at least daily at the same time. An index is computed for each seed lots by dividing the number of normal seedlings removed each day by the day after planting on which they were removed. To take an example, let there be two lots A and B.

$$\text{LOT A} = \frac{\text{No of seedlings removed daily}}{\text{Days after planting}} = \frac{0+0+0+8+10}{1\ 2\ 3\ 4\ 5}$$

2+2=4

$$\text{LOT B} = \frac{\text{No of seedlings removed daily}}{\text{Days after planting}} = \frac{0+0+6+12+25}{1\,2\,3\,4\,5}$$

$$= 2+3+5+=10$$

Here, in this example, we would consider lot B be more vigorous as compared to A which has a lower index.

The main problem in using measurements of rats of germination as a vigour test, lies in standardization, since small difference in the conditions of germination can result in differences in the rate of germination. Careful control of germination conditions is, therefore, necessary.

(3) *Seedling growth rate and seedling dry weight:* Measurements of seedling growth is laboratory or in green house at specific number of day after planting may give an indication of their vigour. The seedling may be further cut and dried at 110 °C for 17 hour to record their dry weight, which will be more for better quality seeds.

(4) *Cool germination test:* This test is generally performed on cottonseeds. Seeds are exposed to constant 18 °C temperature and sufficient humidity to prevent drying of the substrata. Acid delimited cottonseeds are evaluated on sixth day and the machine delinted cottonseed on seventh day. The test studies the effect of cool temperature (18 °C constant) on seed germination and seedling growth rate in cotton.

(5) *Cold test:* Cold test (for Maize) is one of the oldest employed for testing the seed vigour. The test is very useful in estimating the field performance of seed at low temperature soil conditions. Seeds are planted in 2cm thick-leveled soil in a plastic box. The same quantity of soil is then place on top of the seeds, leveled and compacted. Enough cold water (10 °C) is added to the soil to bring the medium to 70% of its water holding capacity. The boxes are then covered and transferred to a refrigerator maintained at 10 °C. After seven day of incubation at 10 °C, the boxes are again transferred to a chamber at 25 °C. Seedling count and/or measurements are made for days after the transfer.

This test was initially developed for corn seed but is equally useful for evaluating soybean, cotton and sorghum seed.

(6) *Paper piercing test:* In this test the seeds are placed on top of ½" thick layer of moist sand and then covered with a special dry filter paper. The paper is again covered with about ¼" of moist sand. The test is done at 20 °C for eight days. This type of test is especially suitable for small grains and cereals. Only the vigorous and healthy seedlings penetrate the paper and come out. The paper used for this test must have the following specifications.: Thickness: 0.4mm, bulk: 4, basic weight 90 gm/m^2, dry bursting strength: 0.3 kg/ cm2 breaking length: 1000-5000mm, filtering speed: 500 ml/min, wet bursting strength :150mm, ash content: 0.1%, fiber composition: chemical wood pulp with high alpha percentage.

(7) *Accelerated ageing test:* This technique requires exposing seeds to 40-45 °C temperatures and 100% relative humidity for 2 to 8 days, followed by a regular germination test. Seed lots that maintain good germination even after accelerated ageing also maintain it under normal storage conditions.

In addition to the routine vigour tests performed in the seed testing laboratories, artificial ageing tests are now increasingly being used. Artificial ageing treatments take advantage of the fact that seed ageing process is determined by the seed moisture level and the temperature. Manipulation of these factors therefore, hastens movement of seeds through the pattern of deterioration. These tests were developed initially by Delouche and Baskin (1973) as a controlled deterioration test. Since then these test have been modified by many other workers depending upon the species and the requirements of experimentation. Accelerated ageing has been recognized as an important vigour test for *Glycine max* which relates to field emergence and storage potential (ISTA, 2003). These tests are usually employed for the following purposes.

- To predict the seed storage potential
- To test the vigour potential of different seed lots
- To study the process of seed deterioration

Principle

The accelerated ageing (AA) stress test exposed seeds for shorter periods to high temperature and high relative humidity (95%). During the test, the seeds absorb moisture from the humid environment and the raised seed moisture content, along with the high temperature causes rapid seed ageing. High vigour seed lots will withstand these extreme stress conditions and age more slowly than low vigour seed lot. Thus, after accelerated ageing, high vigour lots retain a high germination, whilst that of low vigour lot is reduced.

Limitations

- Although the test is particularly useful in picking out seed lots with outstanding high or low vigour, it cannot differentiate clearly between large numbers of seed lots which have similar intermediate vigour levels.
- Selection of appropriate method suitable for particular crop and the accuracy required for attainment of seed moisture content for deterioration requires training and experience in addition to that necessary for routine seed testing.

Sample preparation

Seed sample should be 42-45 grams with moisture content of 10-14%. If the mc is not within the range it should be increased by equilibrating over water or decreased by oven drying at 25 °C. The seeds of known moisture content are placed on moist germination papers and allowed to imbibe water till they attain the desired moisture content (determined periodically). The seeds are then packed and sealed in moisture-proof containers (aluminum foil packets) after excluding the air. The packets are held overnight at 10°C to allow equilibration through the seed lot. After the period of moisture equilibration, the packets are kept in a water bath at 45 °C for 24hr or more depending upon the species after the germination test is performed. The required amount of water to raise the seed moisture to the desired level is added according to the formula given below, in moisture-proof seed packets and held at 10 °C for 24-48hrs.

$$Wf = \frac{Wi\,(100 - Mi)}{(100 - Mf)}$$

where

Wf = Final weight of seeds

Wi = Initial weight of seeds

Mi = Initial seed moisture

Mf = Final seed moisture desired

To maintain seeds at high moisture and temperature, many techniques can be used. A few are mentioned below:

- Seed samples are placed in wire basket held over a measured amount of water in desiccators or any other airtight container and then placed in an incubator at 40-45 °C.
- The seeds are spread on a wire mesh tray and kept over 40ml water in a germination box (11 x 22 x 3.5cm). The box is closed and transferred to an incubator at 40-45 °C.
- After the period of moisture equilibration, the packets are kept in a water bath at 45 °C for 24hrs. (Controlled deterioration)

After which the germination test is performed. In addition, these tests of deterioration can be modified as per the need of experimentations by generating different combinations of moisture and temperature.

Apparatus

Balance, desiccators or plastic boxes, bottle top dispenser, ageing chamber, distilled water, moisture content and germination test facility.

Accelerated ageing procedure

Place 40 ml of the distilled water in each plastic AA box and insert a dry screen tray, being certain not to splash water onto the screen. Weigh a recommended quantity of seed sample and place on the surface of the screen tray. Seeds should be only one layer deep to ensure even uptake of moisture from the humid environment. Place a lid on each plastic AA box. The plastic boxes with seeds should be placed on a shelf from the ageing chamber allowing an air space of approximately 2.5cm between plastic AA boxes to assure temperature uniformity.

Test evaluation

The control and artificially aged seeds are placed on germination substrate using four replicates of 50 seeds each. Appropriate light, moisture and temperature are providing as required in routine germination test. The evaluation is done as per ISAT rules (2003). The emergence of the radical from the seed is taken as an indication of germination.

Precautions

Reproducibility of accelerated ageing results is dependent on precisely controlled conditions. Temperature should be controlled within a tolerance no greater than $\pm$0.50c.Test duration must be strictly adhered to. The longer the test period and less severe accelerated ageing condition, the less critical the test duration become.

The basic assumption of this test is that the germination percentage of a seed lot after a period of accelerated ageing is highly related to the lots vigour and hence to its capacity to perform well under field conditions.

(8) *GADA test:* The glutamic acid decarboxylase activity (GADA) test measures the activity of one specific enzyme. The level of enzyme activity is determined by the amount of carbon dioxide (CO_2) given off and is positively correlated tossed quality; ie., the more the CO_2 given off, the better the seed quality (Grave 19656). Although the GADA test has given valuable result for corn and wheat; it has not proven as valuable for all seeds.

(9) *Tetrazolium test (TZ test):* The test is use more widely to rapidly determine the viability of seed lots, however, it can successfully be used to estimate physiological quality of seed. In this test a chemical 2,3,5-triphenyl tetrazolium chlorides is used as an indicator of sound of weak areas of embryonic tissue. Differences in colour, or lack of colour, along with observation of tissue turgidity or flaccidness, permit recognition of sound, weak and dead tissues.

(10) *Respiration and Respiratory quotient (RQ):* Respiration especially during early hours of imbibitions in water has been shown to be closely correlated with seedling growth rate of lima bean, corn, wheat, soybean and paddy. In corn and paddy, difference in respiration rate have been used to distinguish between high, medium and low vigour seeds. During the process of respiration, oxygen is taken up by the seed and carbon dioxide is evolved. The ratio of volume of carbon dioxide evolved per unit time to the volume of oxygen consumed per unit time is called the respiratory quotient (RQ)

$$RQ = \frac{Q\ CO_2}{Q\ O2}$$

The respiratory quotients were found to be more often related to vigour than was oxygen uptake alone. The arte of gas exchange is measured in the Warburg respirometer.

(11) *Membrane integrity (measurement of leachates):* With ageing the seed membrane becomes more and more permeable. Many substances such as water soluble sugar, free amino & organic acids leach out in the presence of water. The concentration of leachates is measured as electrical conductivity to get an idea of seed vigour.

Measurement of the electrical conductivity of leachates provides an assessment of the extent of electrolyte leakage from plant tissues. Conductivity measurement of the soak water in which a seeds has been steeped gives an estimate of seed vigour. Seed lots that have high electrolyte leakage that is, having high leachate conductivity, are considered as having low vigour, whilst those with low leakage (low conductivity) are considered high vigour.

Factors Affecting Conductivity

1. Ion content of distilled water used for soaking
2. Temperature during soaking of the seed material
3. Length of the soaking period
4. Seed moisture content, and
5. Seed size

Only uninjured seeds should be included in conductivity test, seeds should be randomly drawn from pure seed fraction for this test. AOSA (1983) recommends using untreated seed for conductivity test. However, Loeffler *et al* (1988) found little effect of fungicide seed treatment on conductivity results.

Apparatus Required

1. Conductivity meter
2. Conical flasks /glass beakers
3. Deionised water or distilled water
4. Germinator, incubator or walk in room
5. Moisture content test facilities.
6. Seed sample

Preparation of the sample before measuring conductivity

Determine the moisture content (mc) of the submitted sample according to international Rules for Seed Testing. If the moisture content is below 10% or above 14%, it must be adjusted to between 10 and 14%. To adjust the seed moisture content, mix the fraction of pure seed thoroughly and draw randomly a sub-sample of at least 200 seeds. In the case of a moisture content below 10%, raise the moisture content by placing each weighed sub-sample between moist cloths until it reaches a weight equivalent to between 10% & 14% mc. Experience indicates that to raise the moisture content of pea seeds having an initial moisture content of 7% to a moisture content of 10% or 14% takes approximately 3h and 7h respectively. These times should be taken as a guide only, as the actual times will depend on the extent to which the cloths surrounding the seeds have been moistened. Incase of a moisture content above 14%, reduce the moisture content by placing the weighed sub-sample in an oven at 30 °C until it reaches a weight equivalent to between 10 and 14% mc. Experience indicates that seeds having an initial mc of around 15% takes 1h to reach 14% mc and 5-6h to reach 10% when dried in this way. When the initial seed moisture content is approximately 16%, it takes 1-2h drying to reach 14% mc and 8-10h to reach 10% mc.

The range of seed weights for the sub-samples that will give moisture contents from 10 and 14% can be calculated for both methods using the following equation

$$\text{Weight of sub-sample at 10\% or 14\%} = (\text{initial weight}) \frac{(100\text{-intial mc})}{(100\text{-desired seed mc*})}$$

*The desired mc will be either 10 or14%

When the sub-sample has reached a weight equivalent to between 10 and 14% mc,it should be sealed in moisture -proof container such as an aluminum foil packet or polythene bag and held for 12-18h at 5-10 °C for the moisture content to equilibrate throughout the seed.

Calibrating the dip cell

Prior to use the conductivity meter should be standardized using standard solutions. At least two solutions should be used, one with a conductivity less than 100μcm^{-1} and one with a conductivity between

1000 and 1500μS cm^{-1}. Note that calibrations of the meter using these solutions are carried out at 25 °C which is possible when using a meter with the specifications described above. If the reading is incorrect, the calibration must be repeated and, if necessary, the meter adjusted or repaired. Alternatively, calibrate the conductivity meter using a potassium chloride solution made up in the testing laboratory, if this can be achieved with the accuracy required. Dissolve 0.745 g of pure, dry analytical grade potassium chloride (dried at 150 °C for 1h and cooled in desiccators before weighing) in deionised water to make 1litre of 0.01 M KCl solution. In this solution, the meter should read between 1273 and 1278 μScm^{-1} at 20 °C. If the reading is out of range, the calibration test should be repeated and, if necessary, the meter adjusted or repaired. Conductivity meters that are out of calibration must not be used for the conductivity test.

Conductivity Procedure

Four replicates of 50 seeds are drawn from the pure seed fraction of a seed lot, weighed and placed in 500 ml conical flasks to which 250 ml of deionised water at 20 °C is added. A control flask containing deionised water only is also set up with each test run. The flasks are covered with aluminum foil to decrease evaporation and contamination, and maintained at 20 °C for 24hrs. After removing the foil and gently swirling the solution and seeds in each flask for 10 to 15 s, the conductivity of the soak water is measured by placing a dip cell attached to a conductivity meter, into the solution, taking care to avoid placement directly on to a seed. Several measurements are taken until a stable result is obtained, with swirling repeated between measurements, if required. All the measurements are conducted at 20 °C.

Calculation and expression of results

The conductivity per gram of seed weight for each replicate is calculated after accounting for the background conductivity of the original water, and the average of the four replicates provides the seed lot test results. Thus for each replicate;

$$\frac{\text{Conductivity reading } (\mu S cm^{-1}) - \text{Background reading}}{\text{Weight (g) of replicate}} = \text{Conductivity } (\mu S\, cm^{-1} g\text{-}1)$$

After subtracting the control flask/beaker measurement, conductivity is expressed (µS cm-1g of seed. If replicates differ by more than 5 µS cm^{-1} g-1, the test should be repeated. The difference among the replicates should be within the tolerance ranges e.g. If the reading are from 10-10.9 the maximum tolerance range should be 3.1 (ISTA 2003).

Conclusion

Of all the tests described for evaluating seed vigour, not a single one has proven successful for all kinds of seeds. No single test can satisfy all requirements and a method or combination of methods should be chosen to suit to seed type and sowing environment.

References

Agrawal P.K. and Dadlani M. 1992. Techniques in Seed Science and Technology.

Anon. (1985) International rules for seed testing (Rules+Annexes) Seed Seience & Technol. 13:299-513.

Felfoldi, E.M. (1983). Handbook of pure seed definitions with illustrations (excluding forest tree seeds). ISTA Secretariat, P.O. Box 412, CH-8046 Zurich, Switzerland.

•••

15

Seed Deterioration

Seed deterioration leads to poor seed quality. It may be defined as *"loss of quality, viability and vigor either due to ageing or due to effect of adverse environmental conditions"*. In other words seed deterioration implies *"an irreversible degenerative change in quality of seed after it has reached its maximum quality level"*.

Process of seed deterioration is varied for example short term deterioration in field is likely to be a different physiological event than long term deterioration in storage. Mechanical damage such as lesions influences seed performance differently than uniform physiological damage occurring during seed storage.

The rate of deterioration is influenced by seed moisture content & temperature of storage. In tropical & subtropical conditions, under high ambient relative humidity, seed also deteriorate even when stored under highly controlled condition of gene banks.

Gove 1965 defined deterioration as the *"falling from a higher to a lower level of quality, character or vitality"*. Mineral deficiencies or presence of toxic elements in soil, infestation of a crop by disease or insects & crop damage by heavy rain etc. are some of factors that can adversely affects the seed quality.

Where Does Seed Deterioration Occur?

Seed deterioration does not occur in a set pattern in seeds of different varieties. In monocots dysfunction of sweet corn scutellum & axis has been observed. In wheat TZ staining showed that deterioration begins with root & moves through the embryo in both naturally & artificially aged seeds. Seed deterioration is generally initiated in meristematic areas of seeds & that the radicle tip may be prone to deterioration.

Artificial Ageing

Artificial ageing test are now increasingly being used in seed testing lab as a process by which orthodox seeds are exposed to poor storage condition. Artificial ageing treatment takes advantage of the fact that dead seed ageing process is determined by seed moisture level & temperature. Manipulation of these factors, therefore hastens movement of seed through the pattern of deterioration in seeds.

Artificial ageing test were developed initially by Delouche.& Baskin 1973 & are being employed to test the seed storage potential & vigor potential of different seed lots to study the process of seed deterioration. The ageing process is manifested over a prolonged period during which physiological & biochemical deterioration occurs. As ageing progresses, seeds losses its ability to germinate as well as seed vigor which are reflected as poor field establishment of seeds. These changes are accompanied with delayed or decreased activity of various enzymes involved during germination.

Peroxidases enzyme are required for the generation of reactive oxygen molecules & regulation of hydrogen peroxide level in plants. Peroxides are involved in plant cell wall strengthening by different mechanism like: -

- By cross linking & polymerizing proline rich proteins in cell wall
- By catalyzing lignin deposition.

Lipid per-oxidation is among the most important reason of seed deterioration. It begins with generation of a free radical either by auto-oxidation or enzymatically controlled oxidation. The cells contain complex systems of anti-oxidant defences to protect against harmful effect of activated oxygen molecules which includes enzymes such as peroxidase & catalase. Catalase catalyses the breakdown of hydrogen peroxide to water & molecular oxygen while peroxidase are involved in the oxidation of many organic compounds.

Phosphatases play an important role in phosphorus metabolism during early stages of seeds germination metabolizing phosphate, thus contributing to growth of embryonic axis.

Some of these enzymes only act on certain specified substrates. In such cases it is usually not difficult to assign a physiological function to them. Phosphatase that has extreme broad substrate specificity is generally classified as acid phosphatase. Acid phosphatase catalyse the hydrolysis of a variety of phosphate esters to exhibit pH optima below 6.0 while alkaline phosphatase is active when pH is from 8.0-8.8.

The plant physiologist associated with gene banks and seed storage activity have an unenvitable task of not only providing parameters / indices which reflects the seed quality, but also understanding the basis of deterioration during storage to work out the method for prolonging the shelf life. Therefore, to evolve better storage regimes for *Momordica charantia,* present investigation on physiological & biochemical changes during seed deterioration was under taken in 3 different varieties (*Pant karela-1, Pant karela-2 and Arka harit*).

Physiological Manifestation of Deterioration

Once the seeds have crossed the stage of physiological maturity it moves imperceptivity from one stage of deterioration to another with symptoms appearing in set pattern. As seeds age, they maintain germinability for some time, subsequently they enter a period of rapid decline during which some seeds completely fail to germinate while others germinate & grow normally. The major symptoms observed during deterioration are:-

- Slower growth,
- Delayed germination,
- Changes in seed color,
- Reduced growth of seedlings,
- Increased number of abnormal seedlings,
- Lowered tolerance to adverse storage condition &
- Decreased tolerance to sub-optional environmental condition.

Biochemical Manifestations of Deterioration

Many biochemical changes have been detected in seeds as they deteriorate. These include changes in enzymatic activity, respiratory synthetic pathways, membranes & storage compounds.

1. Changes in respiration

Respiratory metabolism results in oxidation of food reserves, production of a large number of intermediates which serve as building blocks for synthesis of protoplasmic components (proteins, nucleic acid & lipids) & production of biochemical energy in form of ATP for these reactions. Early work suggested that a certain specific amount of heat is produced per unit weight of healthy seeds as a product of respiratory metabolism.

Excess heat indicated infection with microorganism, & subnormal heat suggested low vigor. Subsequent studies on respiration have been carried out by using standard manometric procedures for measuring either oxygen uptake, Carbon dioxide evolution or both.

Changes in respiration during imbibition of deteriorating seeds are generally reflected in lowered rates of oxygen consumption & higher Respiratory Quotient (RQ) values. These changes are pronounced after seed viability declines and are suggested as a possible index of deterioration.

2. Changes in enzymes

Oxidases (catalase, peroxidase & phenolase) were first enzymes investigated with the objective of establishing a co-relation between their activities & seed viability. These enzymes were among those few enzymes which were known & for which simply assay procedures were established. These enzymes play important role in mobilizing & utilizing food reserves during germination. Changes in activity of amylase, oxidase, and decarboxylase have been investigated in deteriorating seed of many crops.

3. Changes in food reserves

The view that loss of seed viability resulted from depletion of food reserves was questioned because when seeds loose their capacity to germinate, they still have a large amount of stored food in their tissues.

Harrington 1967 suggested that depletion of available oxidizable material in meristmatic cell might cause deterioration even when adjacent tissues such as cotyledons & endosperms still contain abundant food. In this case the lack of mobilization of food in dry seed would lead to starvation of these cells. Significant changes in major food reserves take place as seed deteriorate.

•••

16

Estimating Total Soluble Protein

Introduction About Proteins

The seeds were crushed into mortar and pestle so as to obtain a fine powder and were then subjected to defattening solution so that all fats are removed. Process has been repeated several times. The powder is air dried and then was used for extraction of proteins and enzymes.

Preparation of Defattening Solution

- Chloroform : 200 ml
- Acetone : 100 ml
- Methanol : 100 ml

Soluble proteins estimation by Lowry's method

Proteins were extracted using 0.1gms of fine seed powder in 0.01M phosphate buffer (pH 7), precipitated in 10% TCA re-dissolved in 0.1N NaOH and estimated by Lowry's method (Lowry *et al.*, 1951).

Reagents

1. *Solution A:* 2% Na2CO3 (anhydrous) dissolved in 0.1N NaOH.

2. *Solution B:* 0.5% $CuSO4.5H2O$ dissolved in 1% sodium potassium tartrate (freshly prepared)
3. *Solution C:* (Alkaline Copper Sulphate) 50ml of solution A was mixed with 1ml of solution B just prior to use.
4. *Folin-Ciocalteu Reagent (FCR):* Commercial FCR reagent (2N) was diluted with equal volume of distilled water.
5. *Stock Standards Protein Solution:* 50 mg of Bovine Serum Albumin dissolved in 50ml of distilled water (1mg/ml). Various concentrations from 10μg to 100μg protein /ml were made. The standard curve was prepared following the below mentioned methodology.
6. *Working standard protein solution:* 10ml of stock solution was diluted to 50ml distilled water to obtain 200μg protein per ml.

Procedure for protein estimation

Protein forms a colored complex with FCR due to reduction of Phospho-molybdate by Tyrosine & Tryptophan present in protein. Following are the steps for protein estimation:

- Pipetted out 0.2, 0.4, 0.6, 0.8, 1.0ml of working standard solution in a series of test tubes.
- Added 0.1 & 0.2ml of sample extracts in other test tubes. The volume of all test tubes was made up to 1.0 ml with D.W. A tube with 1.0ml D.W. serves as a blank.
- Added 5 ml of solution 'C' mixed well & incubated at room temperature for 10min.
- Added 0.5ml of FCR mixed well immediately & incubated at room temperature in dark for 30min.
- The intensity of blue color was measured colorimetrically at 660nm against the blank.
- Draw a graph & calculate amount of protein in sample, express the result as μg/ml.

Soluble proteins estimation SDS-Poly Acrylamide Gel Electrophoresis

SDS is most widely used method for analyzing protein mixture qualitatively. This experiment was carried out to compare the protein pattern between aged & control seeds.

Preparation of reagents/solutions

a. Tris-HCl buffer (pH 6.6)

- Tris base : 7.2 gm
- Conc. HCl

7.2gm of Tris-base was dissolved in 50ml water, the pH was adjusted with concentrated HCl to 6.6 and then the volume was made up to 100ml with distilled water

b. Stock sample buffer

- 0.6M Tris-HCl buffer (pH 6.6) : 10.4 ml
- SDS : 4.0 gm
- Pyronin-G : 1.0 gm
- Glycerol : 10.0 ml
- Preparation

10.4ml of 0.6M Tris – HCl buffer (pH 6.6) was mixed with 7.9ml water, 4g SDS, 10mg Pyronine-G & 10ml glycerol.

c. Working sample buffer

Working sample buffer solution was made by mixing 4.25ml stock sample buffer solution and 0.75ml Mercaptoethanol and making the volume to 10ml with distilled water.

d. Stacking gel solution (30% Acrylamide solution)

- Acrylamide : 75 g
- Bis acrylamide : 2.0 g
- Preparation

Dissolved in distilled water and the volume made up to 250ml.

e. Separating gel solution (30% Acrylamide Solution)

- Acrylamide : 75 g
- Bis-acrylamide : 1 g

Dissolved in distilled water and the volume made up to 250ml.

f. Separating gel stock buffer (1.875M Tris-HCl Buffer, pH8.8)

- Tris-base : 22.69 g
- Distilled water : 50 ml
- Preparation

Tris-base was dissolved in 50 ml of distil water and the pH was adjusted to 8.8 by HCl and total volume was made upto 100ml with distilled water.

g. Stacking gel stock buffer (0.6M Tris-HCl buffer, pH 6.8)

- Tris-base : 7.26 g
- D.W. : 50 ml

Tris base was dissolved in 50 ml of water and the pH was adjusted to 6.8 by adding HCl and total volume was made upto 100ml with distilled water.

h. Stock SDS Solution (10%)

- SDS : 10 g
- D.W. : 90 ml

i. Ammonium persulphate (5%) (Freshly prepared)

- Ammonium persulphate : 0.5g
- Water to make the volume : 10ml

j. Electrode (Tank) Buffer

- Tris-base : 9.0 g
- SDS : 3.0 g
- Glycine : 42.3 g

These were dissolved in 700ml of D.W. and the volume was made up to 3ltr.

k. Fixing solution (15% TCA)

- Trichloro acetic acid : 150 g
- Water to make the volume 1000 ml

l. Staining solution

- Commasive brilliant blue : 1.25 g
- Methanol : 227 ml
- Acetic acid : 46 ml
- Water to make up volume 500 ml

Or

40 ml of 2% Commassive blue prepared in methanol was added to 400ml of 15 % TCA.

m. Destaining solution

- Acetic acid : 20 ml
- Methanol : 60 ml
- Distilled water : 120 ml

Composition of gel

a. Separating gel

- Tris buffer (pH 8.8) : 12.0 ml
- Distilled water : 7.4 ml
- 30% separating gel acrylamide : 20.0 ml
- 5% APS : 0.4 ml
- 10% SDS : 0.4 ml
- TEMED : 0.04 ml

b. Stacking gel

- Tris buffer : 1.5 ml
- Distilled water : 6.0 ml
- 30% stacking gel acrylamide : 2.0 ml
- 5% APS : 0.4 ml
- 10% SDS : 0.1 ml
- TEMED : 0.04 ml

Methodology

a. Preparation of protein samples

Seeds were crushed with mortar & pestle to fine powder after this powdered material was transferred to test tubes to which 5ml of defattening solvent mixture was added. The tubes were covered with Aluminium foil. After 3hrs the solvent mixture was removed. The process was repeated several times until the oil was removed completely & solution color was transparent. 1000μl of Tris-glycine buffer was added in 0.1gm of deffattened seed powder and the sample was centrifuged at 10,000rpm for 15 min. and thus kept in boiling water bath for 15 min for denaturation. Clear supernatant was quantified by Lowry's method & 30μg of protein was loaded in each well for electrophoresis.

b. Plate assembly

Two glass plates (one notched) were assembled with 2 side spacers.

c. Casting of gel

All the reagents mentioned for separating gel were mixed well and the solution was poured carefully into the vertical cassette leaving an upper margin of 4cms. Care was taken that no air bubble was tapped in the gel solution. The gel surface was overlaid with distilled water using syringes. The gel was left at room temperature for 30 minutes to polymerize. The water layer was poured off after polymerization. After polymerization of running gel the stacking gel mixture poured on the top of the separating gel. An acrylic well forming comb was inserted, ensuring that no air bubble was trapped beneath. The gel was allowed to polymerize for about half an hour. The acrylic comb was carefully removed and the wells were washed with distilled water.

d. Procedure of Electrophoresis

The electrode tank buffer (pH 8.3) was prepared and the lower buffer tank was filled with electrode buffer. The cassette was fitted into the electrophoresis tank. The upper tank was also filled with the electrode buffer. The electrophoresis apparatus was connected to the power pack fitting the electrodes into sockets of the identical color. The voltage & power was turned to maximum. The current was adjusted at a rate of 1.5mA/well. As the tracking dye (pyronin-G) crosses the stacking gel, the current was increased to 2m Amp/well.

The cassette had 13 wells filled with sample, the initial current adjusted was 18.5 mA and the final current was 26 mA. The electrophoresis was stopped just before the tracking dye reached the bottom of gel.

e. Staining & destaining

The gel was stained in 0.1% Commasive brilliant blue for 1 hr & then it was washed with distilled water & destained several times until the band becomes visible.

f. Calculations

The **Rm** value can be calculated by the formula:

$$\textbf{Rm} = \frac{\text{Distance traveled by the band}}{\text{Distance traveled by the tracking dye}}$$

•••

17

Isozyme Analysis

Polyacrylamide gel electrophoresis was used to study the changes in isoenzyme polymorphism in deteriorated seeds. There are many enzymes which are influenced by the process of deterioration but the effects are more pronounced on following enzymes

- Peroxidase,
- Acid phosphatses,
- Esterase and
- Super-oxide Dismutase

a. Extraction of enzyme

10 seeds set for germination in incubators under controlled condition in the dark at 30^{0}C. After 10 days the shoot (without cotyledonary leaves) were taken for the study of enzyme activity. The material was grinded in 0.5ml of Tris-buffer (pH 7.4) and then centrifuged at 10,000rpm for 15min. The supernatant was used for loading after adding dye Bromophenol blue (prepared in sucrose)

b. Reagents for enzyme extraction

Tracking dye (Bromophenol blue 10.5% w/v)

1. Bromophenol blue : 25 mg
2. Tris Hcl buffer : 10 ml
3. Sucrose solution (30%) : 5 ml

Extraction buffer (Tris buffer (pH 7.4))

1. 1.21gm Tris buffer : 100 ml D H20
2. 8gm Sucrose in : 100 ml D. H2O
3. 0.01gm Ascorbic acid in : 10 ml D. H20
4. 0.01gm Cystine HCl in : 10ml D. H20

The entire reagents mixed & pH adjusted to 7.4 with 1N HCl

c. Reagents for enzyme extraction:

- Tris buffer (pH 8.8) : 30 ml
- Distilled water : 18.5 ml
- 30% running gel acrylamide : 50 ml
- 5% APS : 10 ml
- 10% SDS : 10 ml
- TEMED : 0.04 ml
- Preparation of gel

The gel plates were cleaned and dried. The cassette was assembled using the gel casting stand. These were mixed thoroughly and gel solution was poured into the cassette. An acrylic well forming comb was inserted and it was allowed to polymerize for 1 hr. The comb was removed and wells were rinsed with distilled water. The samples were loaded using syringe.

d. Electrophoresis for Isozyme

The gel cassette was fitted to the apparatus and was fitted with electrode buffer. Voltage and power were fixed at 150 and was run for 4 hr at 10^{0}C. After this the gel were stained for different isoenzymes.

e. Staining solutions and Procedure

Peroxidase enzyme

Stain composition

O-Dianisidine : 16 ml

Preparation: 0.5 gm of O-Dianisidine was dissolved in 100 ml of 0.1 M HCl (prepared by 8 ml of HCL was added to 100 ml of D.W.)

Sodium Acetate buffer (pH-5.4) : 48 ml

Preparation: 0.6 M Sodium acetate was prepared by adding 8.16 gm of Sodium acetate in 100 ml of D.W. and 0.6M Acetic acid solution prepared by dissolving 2.5 ml of Acetic acid in 50 ml D.W. 43 ml of 0.6 M Sodium acetate solution was taken & added to 7ml of 0.6M Acetic acid.

Distilled water : 416 ml

H_2O_2 : 3 %

Stain preparation

Take 48 ml 0.6 M Na Acetate buffer (pH-5.4) solution and add 416 ml of distilled water and 16 ml O-Dianisidine solution to it. This will give us 500ml of stain solution. When stored in refrigerator, this stock solution was stable for a month.

Staining procedure

Incubate gel in the staining solution for 15 minutes, the solution was removed and 3 % H2O2 solution was added. The bands appeared green initially and then turned brick red color. The enzyme activity was terminated by adding 7 % Acetic Acid as fixing solution.

Acid phosphates

Staining solution

1. Solution A: (0.2 M Sodium acetate buffer)
   - 0.2 M CH_3COOH : 1.155 ml of CH_3COOH in 100 ml of D.W.
   - 0.2 M sodium acetate : 2.22 gm CH_3COONa in 100 ml of D.W.

   Mix 41ml of 0.2 M CH_3COOH and 59ml 0.2 M sodium acetate to obtain 0.2 M Sodium acetate buffer of the $_pH$ of 4.8.
2. Solution B

- 0.2 M Sodium acetate buffer (pH 4.8) : 20ml
- a-Napthyl acid phosphate : 20mg
- Fast blue RR : 20mg
- NaCl : 0.4g
- $MgCl_2$ (10%) : 2-4drops

Staining procedure

The gel was placed in Solution A for ½ an hour and then the gel was transferred in Solution B overnight to get clear banding patterns.

Super Oxide Dismutase

Staining solution

Solution A

- 0.05 M Na Phosphate (pH 7.5) : 100 ml
- MTT : 1 ml (10 mg/ml)

Solution B

- 0.05 M Na Phosphate (pH 7.5) : 100 ml
- TEMED : 0.2 ml
- Riboflavin : 1 mg

Staining procedure

The gel was soaked in dark in solution A at 37 °C for 20-30 minutes. Solution A was decanted and Solution B was added followed by incubation of the gel under illuminator for 15 minutes or until achromatic bands appear. The bands were then fixed in 40 % glycerol for about 45 minutes. The reaction was stopped by washing the gel with D.W.

Esterase

Staining solution

- 0.1M Tris-HCl buffer (pH 7.2) : 100 ml
- a-Naphthyl acetate (phosphate) : 40 mg
- Fast Blue RR : 100 mg
- Acetone : 5 ml

Staining procedure

Incubate the gel in dark at 37°C in 100ml of 0.1M Tris-HCl buffer (pH 7.2) containing 40 mg a-Naphthyl acetate (dissolved in 5 ml acetone) and 100mg of Fast Blue RR salt. Reddish brown bands appear for esterase. Fix the gel in 7% acetic acid solution.

•••

18

DNA Analysis

Introduction about the changes in DNA levels

This experiment completed following 6 steps:

1. DNA extraction
2. Purity testing of DNA in samples
3. DNA purification
4. DNA quantification
5. Polymerase chain reaction
6. Gel electrophoresis

1. DNA extraction using CTAB method

1.1 Reagents

- CTAB buffer (Cethyl trimethyl ammonium bromine)

  1.5% CTAB = 7.5 ml of 20% CTAB stock

  100mM Tris = 10 ml of 1M Tris stock

  20mM EDTA = 4 ml of 0.5M EDTA stock

1.4M NaCl	=	46.6 ml of 3M NaCl stock
D.W.	=	31.9 ml
Final volume	=	**100 ml**

Adjust pH to 8.0 with HCl.

- TE buffer (Tris-EDTA buffer of pH 8.0)

10mM Tris	=	1 ml of 1M Tris (pH 8.0) stock
1mM EDTA	=	0.2 ml of 0.5M EDTA (pH 8.0) stock
Sterilized D.W.	=	98.8 ml
Final volume	=	**100 ml**

Autoclaved before use this buffer.

- Liquid Nitrogen
- 2% 2- Merceptoethanol
- 0.5% Polyvinyl pyrollidine
- Chloroform : Iso-amylalcohol mixture (24:1)
- Iso-propanol
- Sodium acetate
- 70% ethanol

1.2 Procedure

Steps for DNA extraction are as follows:

1.2.1. 5gm fresh leaves from each sample were ground in to fine powder using a pastle and mortar in apool of liquid nitrogen in laminar air flow chamber

1.2.2. Powdered material of each sample transferred to 50ml centrifuged tubes with help of spatula. To this 20 ml CTAB buffer (prewarmed at 60°C), 0.4 ml of 2-Merceptoethanol and 1/10th volume of PVP was added. 2-Merceptoethanol (reducing agent) and EDTA (chelating agent) helps to inactivate nucleases that are released from plant cell an can cause degradation of genomic DNA. Phenolic compounds may be released on disruption of plant tissues and these may interfere with subsequent usus of the DNA such as in PCR. Phenolic compounds can be removed by adding PVP to the buffer.

1.2.3. Mix the contents of tubes properly and kept in water bath at 60°C for 1hr with periodically inverting.

1.2.4. *20 ml of Chloroform :* Isoamylalcohol mixture was added to each centrifuge tube and tube gently swirled for 10min. This helped in separation of Proteins and polysaccharides.

1.2.5. The tubes were centrifuged at 13000-14000 rpm for 10-15min at 20-25°C and the supernatant (clear solution) collected carefully.

1.2.6. *Now add Iso :* Propanol (2/3rd the volume of supernatant) and Sod. Acetate (1/10th the volume of supernatant) to the supernatant in each tube. The tubes were slowly tilted for a few minutes and kept in the refrigerator overnight. This will leads to the precipitation of DNA.

1.2.7. The tubes were centrifuged at 13000-14000 rpm for 20min at 20-25°C and the pellets collected carefully.

1.2.8. Wash DNA pellets with 70% ethanol (freshly prepared).

1.2.9. 70% ethanol was added to the tubes and these tubes were centrifuged at 13000-14000 rpm for 20min at 20-25°C and the pellets collected carefully. The pellets allowed to dry nside the tubes for few hrs under vacuum.

1.2.10. Dissolve the DNA pellets in minimum volume of TE buffer and stored at -20°C.

2. Purity testing of DNA in samples

To check the presence of DNA in samples horizontal Agarose gel electrophoresis is done.

2.1 Reagents

- 0.5 X Tris Borate EDTA (TBE) buffer

It can be prepared by diluting the 10X stock TBE buffer whose composition is as follows:

Tris	=	10.8 gm
0.5M EDTA	=	4 ml
Boric acid	=	5.5 gm

D.W. to make up final volume up to 100ml.

- Loading dye : -

0.2% Bromophenol blue	=	2 µl
0.2% Xylene cyanol	=	2 µl
60% glycerol	=	600 µl
60mM EDTA	=	120 µl
D.W.	=	276 µl
Final volume	=	**1000 µl**

- Agarose powder
- Ethidium Bromide (5 µl/100ml)
- 0.5M TBE buffer

2.2 Procedure

2.2.1. 800 gm of agarose was added to 100 ml of 0.5X TBE buffer and the mixture was boiled till complete dissolution of agaroge and gently cooed to 50-55°C on a shaker.

2.2.2. Ethidium bromide was added to gel and this mixture was poured over the horizontal gel casting assembly. The comb was also put in to position carefully.

2.2.3. The comb was removed after solidification of gel.

2.2.4. Solidified gel transferred to electrophoresis tank (filled with 0.5M TBE buffer) and pre-run for 2-3 hrs to allow ethidium bromide to move through the gel. This facilitated visibility od DNA under UV transilluminator because ethidium bromide is an intercalating agent which enters in DNA base pairs and make them visible.

2.2.5. 2 µl of loading dye, 6 µl sterilized water and 2 µl of dissolved genomic DNA mixed and loaded in the wells on the gel. The gel was run at 65 V for ½ an hr and after some time voltage was increased up to 80 V.

2.2.6. At completion of electrophoresis DNA was visualized under ultraviolet light of 230nm wavelength.

3. DNA purification

When DNA was visualized under UV light a bright smear represents RNA contamination which was removed by purifying DNA as follows:

3.1 Reagents

- RNA ase solution (10 mg/ ml)

10mM Tris (pH 7.5)	=	10 µl of 1M stock
15mM NaCl	=	15 µl of 1M stock
RNA ase	=	10 mg
D.W.	=	975 µl
Final volume	=	**1 ml**

The RNA ase solution was heated at 100°C for 15 min cooled to room temperature and stored at -20°C

- TE buffer (Tris-EDTA buffer of pH 8.0): -
- Phenol : Chloroform : Iso-amylalcohol mixture (25:24:1)
- Chloroform : Iso-amylalcohol mixture (24:1)
- Iso-propanol
- Sodium acetate
- 70% ethanol

3.2 Procedure

3.2.1 5 µl of RNA ase solution was added to 1 ml of DNA sample and mixture was incubated at 37°C for 40 min and an equal volume of Phenol : Chloroform : Isoamylalcohol mixture was added. The tubes were shaken well for few min.

3.2.2 The tubes were centrifuged at 13000-14000 rpm for 10-15min at 20-25°C and the supernatant (clear solution) collected carefully.

3.2.3 Equal volume (to supernatant) of Chloroform : Isoamylalcohol was added to tubes, shaken wall and tubes were centrifuged at 13000-14000 rpm for 10-15min at 20-25°C.

3.2.4 Again supernatant (clear solution) was collected. To this 3/5th volume of Iso- propanol and 1/10th volume of Sod. Acetate was added and tubes were kept overnight at -20°C. This facilitated the precipitation of DNA.

3.2.5 The tubes were centrifuged at 13000 rpm for 15min at 20°C and the supernatant discarded. The DNA pellets in each tubes was washed in 70% Ethanol, then again immersed in fresh 70% ethanol and centrifuged at 13000 for 8-15min at 25°C.

3.2.6 After pouring out ethanol the pellets were air dried for 2-3 hrs, dissolved in 500 µl TE buffer and kept at -20°C or -80°C for long term storage.

4. DNA quantification

It was done to know amount of DNA present in purified samples. Quantification was done by using flourimeter (Hoefer's). Flourimeter measures the fluorescence emitted by double stranded DNA-Hoechst 33258 dye complex which is directly proportional to amount of DNA in sample. Hoechst 33258 dye bind only with DNA but not with RNA.

4.1 Reagents:

- 10 X TNE buffer

1M Tris	=	100 ml
0.5M EDTA	=	20 ml
3M NaCl	=	666.6 ml
D.W.	=	213..4 ml
Final volume	=	**1000 ml**

  Adjust ph 7.4 by HCl , filter and autoclave before use, store at 4°C for 3 months.

- Hoechst 33258 dye stock (10 mg / 10 ml of D.W.) was stored at 4°C in an amber bottle.

- Assay solutions

  i. Assay solution A (for low DNA conc. = 10-500 ng / ml)

Hoechst 33258 stock solution	:	10 μl
10 X TNE buffer	:	10 ml
Filtered D.W.	:	90 ml

ii. Assay solution B (for high DNA conc. = 100-5000 ng / ml)

Hoechst 33258 stock solution	:	100 μl
10 X TNE buffer	:	10 ml
Filtered D.W.	:	90 ml

- **DNA standard solution** using calf thymus DNA was prepared as under

1 mg/ml of DNA standard stock	:	100 μl
10 X TNE buffer	:	100 μl
Filtered D.W.	:	800 μl

4.2 Procedure

4.2.1. Fluorimeter was turned on at least 15 min before using.

4.2.2. Zero the instruments: Prepare an assay blank using 2 ml assay appropriate assay solution (assay solution A or B). Dry the sides of cuvette and insert it in to well, close the lid and press "ZERO". After 'zero' displays remove cuvette.

4.2.3. Calibrate the instrument: Deliver 2μl of standard DNA solution (low or high conc.) to 2 ml of assay solution in the cuvette, mix it properly and placed in well. Press "CALIB". Enter 100 for low range assay and 1000 for high range and than press "ENTER". After the entered value displayes remove the cuvette.

4.2.4. Zero the instrument: Empty and rinst cuvette. Add 2 ml assay solution insert cuvette in well and press "ZERO".

4.2.5. Measure the sample.

5. Polymerase Chain reaction

5.1 Reagents

- **Reaction mixture**

0.5mM MgCl2	=	0.5 μl of 25mM stock

0.2mM dNTPs mixture	=	2.5 µl of 2mM each of dATPs, dCTPs, dGTPs, & dTTPs
0.4 µM Primer*	=	1 µl of 10 µM stock
1 X buffer	=	2.5 µl of 10 X stock
Taq polymerase (3U/µl)	=	2.5 µl
Sterilized D.W.	=	13.25 µl

To each tube was added 5.0µl (25ng) of genomic DNA to make a total volume of 25µl per tube.

5.2 Procedure

5.2.1. The tubes were centrifuged for a min. at 6000-8000rpm and placed in the PCR machine. The initial denaturation step was carried out at 94°C for 4min followed by 40-50 cycles with the following cycle parameters.

Step 1. 94°C for 1 min.

Step 2. 35°C for 1 min.

Step 3. 72°C for 1 min.

The reaction was run for 4 ½ hr.

6. Gel electrophoresis

When amplification was finished 3µl of loading dye was added to each sample and the control. The DNA was now loaded on to a 1.2% agarose gel in 0.5X TBE buffer with ethidium bromide. The gel was run in the buffer at 55V for 4-5 hr, visualized on a UV transilluminator and photographed using Polaroid 665 or 667 film.

•••

19

Recent Developments Concerning Biodiversity

Plant Genetic Resources for Food and Agriculture (PGRFA)

Plant genetic resources (PGR) are one of the important components of natural resources and hold the key to solid foundation of agriculture and food and nutrition security. The last three decades of the 20th century, witnessed an era, wherein agriculture registered rapid progress and along with that, the collection, conservation and sustainable utilization of PGR gained importance.

The need to preserve, exchange and utilize plant genetic resources is now recognized worldwide. Even countries with great genetic diversity in certain crops are heavily dependent on many other crops introduced from other areas. As an indicator of agro biodiversity interdependence, the production accounted for by non native crops is 100 % in Australia and North America but only 31-34% in West central Asia and Indo-China respectively. The history of the use of dwarfing genes in wheat provides an excellent example of the benefits of unrestricted exchange of genetic resources. American wheat varieties were used in Japan in the breeding of Norin 10, which was then used in the USA to breed the dwarf and semi-dwarf varieties used subsequently as parents in most wheat producing countries including Japan. The benefits of this free exchange of germplasm have been enjoyed by wheat growers and consumers world- wide. International exchange of germplasm resources has promoted crop improvement and agriculture development in various countries. All the rust resistant wheat varieties used in china have been derived from foreign varieties. In recent times however, questions are being asked regarding sharing of these resources. Will seeds be freely available from all collections? Will germplasm rich nations cut off germplasm supply and claims be made by developed nations without proper economic considerations, especially by the multinational seeds companies? A great number of accessions in genebank are primitive cultivars and landraces from peasant agriculture; therefore they should be paid for this germplasm.

Three successive international negotiation processes impacted on this rise of PGR to prominence. These are the negotiations of the Convention on Biological Diversity, the Agreement on Trade-Related Aspects of Intellectual Property Rights as part of the Marrakesh Agreement establishing the World Trade Organization and the International Treaty on Plant Genetic Resources for Food and Agriculture. These negotiations, in conjunction with rapid advancements in the fields of molecular biology, biotechnology and bioinformatics, have led to the emergence of new legal, political and technological regimes that dictate PGR management. The discussions on ownership and utilization of germplasm and benefits therefore, and related issues resulted in a paradigm shift from PGR being a 'common heritage of mankind' to 'sovereign rights of nations'.

Earlier PGR were used to be considered as 'common heritage of humankind and access to PGR under that regime was fairly liberal for public and private sectors across the countries and International Agricultural Research Centers (IARC). Advances in science and technology have significantly influenced and changed the course of utilization vis-à-vis valuation of the PGR. New international conventions heralded the era of property rights, both intellectual and sovereign. The intersection between the intellectual property and PGR is giving rise to an uncertainty. At this juncture, need was realized for developing a clearer understanding of the paradigm shift, rationalizing the challenges, revisiting the action plan, and refining the implementation strategy.

With the adoption of Convention of Biological Diversity (CBD) in 1992 the commonly accepted ideology of "common heritage" for plant genetic resources came under pressure. The Convention on Biological Diversity will alter procedures and policies for germplasm exchange. The obligations arising from the CBD that subject access and use of genetic resources to prior informed consent and mutually agreed terms on bilateral basis have rendered the process of access and freedom to operate very cumbersome and unpredictable. Exchange of genetic resources has reduced considerably, except for the pre-CBD accessions in the gene banks. The multilateral system (MLS) included in the FAO International treaty on Plant Genetic Resources in Food and Agriculture will facilitate accesses to genera and species included in Annex 1 of the treaty. It should bring stability and predictability to germplasm transfer of those crops.

Plant breeders have been among the first, led by the pioneer Vavilov, to have drawn attention to the importance of genetic resources and their preservation. Activities on PGRs go back as far as 1947 when FAO put collection and preservation of PGR on its agenda. Eucarpia drew attention to this important subject as early as 1962.But preserving genetic resources is not enough. Plant breeders must have access to them for further research and breeding. Until recently this was the case for the two classes of genetic resources described below.

As every plant breeder is aware, diverse gene pools are the foundation of an effective crop improvement programme. Not only do new gene pools provide the necessary building blocks for further varietal improvement, but also genetic diversity is essential if high levels of productivity are to be sustained. Plant Genetic Resources (PGR) can be dived into two broad categories:

- Wild relatives and landraces (possibly including obsolete varieties) that may be considered as a gene reservoir.
- Modern commercial varieties that may be considered as the "building blocks" on which the incremental progress in plant breeding is built.

Plant Genetic Resources were considered in the past as "common heritage of mankind to be preserved and freely available for use, for the benefit of present and future generations" as per the words used by the FAO Conferences of November 1983 when adopting the

International Undertaking (IU) on PGR. However, the ideology of "common heritage" began to come under pressure in the 1980s due to a perceived imbalance in the legal regimes applicable to genetic resources: the development of patents on biotechnological inventions on the one hand and the free access to landraces and wild relatives on the other. A number of other factors put additional pressure on the "common heritage" approach.

World Trade Organization (WTO)

A parallel development under the World Trade Organizaiton which had a bearing on the PGR was the Article 27.3 (b) of the Trade Related Aspects of Intellectual Property Rights (TRIPS), which made it obligatory for the member country to "provide protection to plant varieties either by patent or by an effective sui generis system or by any combination thereof'. The developed counties generally had a well-established mechanism of plant patents or plant variety protection such a Union International pour la Protection des abstentions Vegetables (UPOV, International Union for protection of new varieties of plants) and it became obligatory for the developing countries also to have some system of protection of plant varieties.

The FAO International Treaty on Plant Genetic Resources for Food and Agriculture ITPGR(FA)

The Food and Agriculture Organization (FAO) of United Nations established an inter-governmental Commission on Genetic Resources for Food and Agriculture (CGRFA) in 1983 which adopted a non-binding International Undertaking on Plant Genetic Resources (IUPGR), aimed at promoting the conservation and use of PGR, ensuring access to them without undue restriction with the aim that PGR would remain available in public domain. The IUPGR provided for the germplasm exploration, collection, *in situ* and *ex situ* conservation. It also provided for availability of germplasm and international cooperation for plant breeding, *ex situ* conservation and information systems.

The 1993 FAO Conference provided the framework for the revision of the IUPGR in harmony with the CBD and negotiations for the revision of IUPGR began at the First Extraordinary Session of the CGRFA in November 1994. These continued for seven years until the adoption of the International Treaty on Plant Genetic Resources for

Food and Agriculture (ITPGRFA) in November 2001 at the 31st Conference of the FAO.

The ITPGRFA became effective from 29 June 2004 after ratification by 40 countries. Presently the Treaty has 115 Contracting Parties. Each country that ratifies will have to develop the legislation and regulations it needs to implement the Treaty. The Secretariat of ITPGRFA has also requested the Contracting Parties to declare their respective focal points. Forty eight Parties have declared their National Focal Points as on 23.10 2007. India signed and ratified the Treaty on 10 June 2002. However, the legislation and regulations needed to implement the Treaty by the country are yet to be affirmed or developed. India is also yet to officially declare its national focal point for the implementation of ITPGRFA.

The Biological Diversity Act, 2002 of India has provisions in its Chapter 2 governing access to bio-resources to foreign nationals. However, there is no provision in the Act that may expressly address the agreed provisions of multilateral access and benefit sharing under the Treaty. The Department of Agriculture and Cooperation (DAC), Government of India is the nodal agency for the implementation of the Treaty in the country. The National Bureau of Plant Genetic Resources (NBPGR), a constituent of the Indian Council of Agricultural Research (ICAR), being mandated with the conservation and sustainable use of PGRFA, is duly interested and concerned for the effective implementation of the relevant provisions of the Treaty in India.

Objectives of the Treaty

The objectives of the Treaty are 'the conservation and sustainable use of plant genetic resources for food and agriculture and the fair and equitable sharing of benefits arising out of their use, in harmony with the Convention on Biological Diversity, for sustainable agriculture and food security'.

Obligations on part of Contracting Parties

Each contracting party of the Treaty shall ensure the conformity of its laws, regulations and programmes with the provisions of the Treaty. Each contracting party shall, subject to national legislation, and in cooperation with other contracted parties where appropriate, promote an integrated approach to the exploration, conservation and sustainable use of PGRFA and shall in particular, as appropriate, undertake some of the activities like: Survey and inventory of PGRFA; Promote the collection of PGRFA and relevant associated information on those PGR that are under threat or are of potential use; Promote or support, as appropriate, farmers and local communities' efforts to manage and conserve on-farm their PGRFA; Promote *in situ* conservation of wild crop relatives and wild plants for food production; Cooperate to promote the development of an efficient and sustainable system of *ex situ* conservation, giving due attention to the need for adequate documentation, characterization, regeneration and evaluation; Take steps to minimize or, if possible, eliminate threats to PGRFA.

The contracting parties are also obligated to develop and maintain appropriate policies and legal measures to promote sustainable use of PGR for sustainable agriculture and food security, and international cooperation, in the following manner, at national level.

(i) Development and maintenance of diverse farming systems;

(ii) Strengthening research which enhances and conserves biological diversity by maximizing intra- and inter-specific variation;

(iii) Promoting breeding efforts for development of varieties adapted to social, economic and ecological conditions;

(iv) Broadening genetic base of crops;

(v) Promoting locally adapted crops and varieties;

(vi) Supporting the wider use of diversity of varieties and species in on-farm management;

(vii) Reviewing policies for varieties release and seed production as appropriate.

At the international level, countries are required to cooperate with other contracting parties, directly or through FAO and the relevant international organizations in the conservation and sustainable of PGRFA and also cooperate in establishment or strengthening the capabilities of developing countries with respect to conservation and sustainable use of PGRFA.

The Multilateral System of Access and Benefit Sharing

In exercise of sovereign rights of nations over their PGRFA, the Treaty envisages the establishment of a multilateral system (MS) covering a list of 35 food and 29 forage crops (Annexure 1 of the Treaty) selected on the criteria of food security and inter-dependence. It includes all PGRFA of the listed crops that are under the management and control of the contracting parties and in public domain. The MS shall also include the PGRFA as listed, and held in the *ex situ* collections of the International Agricultural Research Centers of the Consultative Group on International Agricultural Research (CGIAR).

The access to PGRFA shall be provided solely for the purpose of utilization and conservation for research, breeding and training for food and agriculture, provided that such purpose does not include chemical, pharmaceutical and/or other non-food/feed industrial uses [Article 12.3 (a)]. The access is to be accorded expeditiously without the need to track individual accessions and either no fee or fee, not exceeding the minimal cost involved, is to be charged. It is also envisaged in the Treaty that all the available passport data, subject to applicable law, and any other associated available non-confidential descriptive information should be made available with PGRFA being provided. However, as per Article 12.3 (d), the recipient shall not claim any intellectual property or other rights that limit the facilitated access to PGRFA, or their genetic parts or components, in the form received from MS. The MS also makes it mandatory [Article 12.3 (g)] on the part of recipient to continue to make the accessed and conserved material available to MS.

Article 13 of the Treaty describes about the benefit sharing mechanisms in the MS. It has been indicated that the benefits arising from the use, including the commercialization of PGRFA under MS shall be shared fairly and equitably through different mechanisms like the exchange of information, access to and transfer of technology, capacity building, and the sharing of the benefits arising from commercialization, taking into account the priority activity areas in the rolling Global Plan of Action, under the guidance of the Governing Body.

Article 13.2 d (ii) of the Treaty states that "a recipient who commercializes a product that is a PGRFA and that incorporates material accessed from MS, shall pay to the mechanism, such as Trust Account, as referred in the Article 19.3 (f) of the Treaty, an equitable share of the benefits arising from commercialization of that product, except whenever such a product is available without restriction to others for further research and breeding, in which case the recipient who commercializes shall be encouraged to make such payment".

Protection of Plant varieties and Farmers' Right Act

The Protection of Plant Varieties and Farmer' Right Act (PPVFR), 2001, of India provides for protection of plant varieties and and farmer rights. As per the Act plant variety is a plant grouping, except microorganism, within a single taxon of the lowest rank, which can be defined by the expression of the characteristics resulting from a given genotype of that grouping or distinguished from any other grouping by expression of at least one characteristic. A new variety shall be registered if it conforms to the criteria of novelty, distinctiveness, uniformity and stability. Each application for registration shall contain among other information a complete passport data of the parental lines from which the variety has been derived along has been taken and all such information relating to the contribution, if any, of any farmer, village community, institution or organization in developing the variety. The registration is to be valid for 9 years in case of tree and vines and 6 years incase of other crops and may be reviewed for 9 more years. In case of extant varieties the protection is to be for 15 years form the date of notification under the Seed ACT, 1966. Under the provision of farmers Rights, any community of a civil society organization representing them can file a claim for benefit sharing regarding the contribution of people to a variety granted protection. After verification, for the establishment of a National gene fund to which funds from benefit sharing, compensation, annual fee by way of royalty and contribution from any national and international organization and other sources will be credited. The Gene Fund shall be used to meet the requirements of amount to be paid by way of benefit sharing and compensation, expenditure for supporting the conservation and sustainable use of genetic resources including in sity any ex situ collections and for strengthening the capability.

Biological Diversity Act

The Biological Diversity Act (BDA), 2002 of India, is a broad policy statement identifying the strategies and the actions that need to be taken for access to genetic resources, their conservation, sustainable use the fair and equitable sharing of benefits as per the provisions of the CBD. The Act envisages establishing a National Biodiversity Authority, State Biodiversity Boards and local level biodiversity management committees, to provide for regulated access to bioresearches and the traditional knowledge associated with them. It makes such access subject to terms and conditions that secure equitable sharing of benefits for the resources accessed. It provides guidelines for undertaking collaborative research involving the use of bio resources and associated knowledge. The Act also provides for protection from any patent in or outside India on any invention or information on bio resources obtained from India without prior approval of the National Biodiversity Authority which shall impose a benefit sharing fee or royalty or both. A National Biodiversity Fund shall be created with the Authority that shall be applied for channeling benefits to the claimers, conservation and promotion of bio resources and development of areas form where such bio resources or knowledge associated thereto has been accessed, and socio-economic development of areas in consultation with local bodies concerned. The National Biodiversity Authority with its headquarters at Chennai deals with biological diversity of agriculture-related species and their wild relatives.

ANNEXURE I

List of crops covered under the multilateral system

FOOD CROPS		
Crop	**Genus**	**Observations**
Breadfruit	Artocarpus	Breadfruit only.
Asparagus	*Asparagus*	
Oat	*Avena*	
Beet	*Beta*	
Brassica complex	*Brassica* et al.	Genera included are: *Brassica, Armoracia, Barbarea, Camelina, Crambe, Diplotaxis, Eruca, Isatis, Lepidium, Raphano-brassica, Raphanus, Rorippa,* and *Sinapis*. This comprises oilseed and vegetable crops such as cabbage, rapeseed, mustard, cress, rocket, radish, and turnip. The species *Lepidium meyenii* (maca) is excluded.

Contd...

Contd...

Pigeon Pea	*Cajanus*	
Chickpea	*Cicer*	
Citrus	*Citrus*	Genera *Poncirus* and *Fortunella* are included as root stock.
Coconut	*Cocos*	
Major aroids	*Colocasia, Xanthosoma*	Major aroids include taro, cocoyam, dasheen & tannia.
Carrot	*Daucus*	
Yams	*Dioscorea*	
Finger Millet	*Eleusine*	
Strawberry	*Fragaria*	
Sunflower	*Helianthus*	
Barley	*Hordeum*	
Sweet Potato	*Ipomoea*	
Grass pea	*Lathyrus*	
Lentil	*Lens*	
Apple	*Malus*	
Cassava	*Manihot*	*Manihot esculenta* only.
Banana/Plantain	*Musa*	Except *Musa textilis*.
Rice	*Oryza*	
Pearl Millet	*Pennisetum*	
Beans	*Phaseolus*	Except *Phaseolus polyanthus*.
Pea	*Pisum*	
Rye	*Secale*	
Potato	*Solanum*	Section tuberosa included, except *Solanum phureja*.
Eggplant	*Solanum*	Section melongena included.
Sorghum	*Sorghum*	
Triticale	*Triticosecale*	

Contd...

Contd...

Wheat	*Triticum* et al.	Including *Agropyron, Elymus,* and *Secale.*
Faba Bean/Vetch	*Vicia*	
Cowpea et al.	*Vigna*	
Maize	*Zea*	Excluding *Zea perennis, Zea diploperennis,* and *Zea luxurians.*

Forages

Genera	Species
LEGUME FORAGES	
Astragalus	*chinensis, cicer, arenarius*
Canavalia	*ensiformis*
Coronilla	*varia*
Hedysarum	*coronarium*
Lathyrus	*cicera, ciliolatus, hirsutus, ochrus, odoratus, sativus*
Lespedeza	*cuneata, striata, stipulacea*
Lotus	*corniculatus, subbiflorus, uliginosus*
Lupinus	*albus, angustifolius, luteus*
Medicago	*arborea, falcata, sativa, scutellata, rigidula, truncatula*
Melilotus	*albus, officinalis*
Onobrychis	*viciifolia*
Ornithopus	*sativus*
Prosopis	*affinis, alba, chilensis, nigra, pallida*
Pueraria	*phaseoloides*
Trifolium	*alexandrinum, alpestre, ambiguum, angustifolium, arvense, agrocicerum, hybridum, incarnatum, pratense, repens, resupinatum, rueppellianum, semipilosum, subterraneum, vesiculosum*
GRASS FORAGES	
Andropogon	*gayanus*
Agropyron	*cristatum, desertorum*
Agrostis	*stolonifera, tenuis*

Contd...

Contd...

Alopecurus	*pratensis*
Arrhenatherum	*elatius*
Dactylis	*glomerata*
Festuca	*arundinacea, gigantea, heterophylla, ovina, pratensis, rubra*
Lolium	*hybridum, multiflorum, perenne, rigidum, temulentum*
Phalaris	*aquatica, arundinacea*
Phleum	*pratense*
Poa	*alpina, annua, pratensis*
Tripsacum	*laxum*
OTHER FORAGES	
Atriplex	*halimus, nummularia*
Salsola	*vermiculata*

•••

20

The New Seed Bill 2004

A good harvest is impossible without a good seed. For agriculture to prosper, farmers must have a reliable supply of high quality seed and seedlings, at an affordable price. Sustained increase in agriculture production and productivity is dependent, to a large extent, on timely supply of the quality seeds of new varieties and hybrids, in required quantities to the farmers will help in faster dissemination and adoption of the new varieties/hybrids and new technologies by the farmers. The quality of the seed may get affected during its production to marketing. Therefore it is necessary to ensure the quality of seed till it reaches the farmers through legal methods, *viz.* Seed certification and quality control mechanism.

To control the quality of seed during production and marketing, Government of India (GOI) had passed the **Seed Act (1966)** and farmed the **Seed Rules (1968).** With the advances in Indian seed industry, **Seed Control Order (1983)** was passed by GOI to keep the check on the persons involved in seed business through the compulsory licensing. In order to make available the seed of improved varieties to the Indian farmers, government has relaxed the restriction on importing of seed materials from foreign countries by enacting the **New Seed Policy (1988).** For the effective control of seed quality, GOI has come out with a revised **Seed Bill 2004,** amalgamating and overcoming the shortcomings of provisions in the earlier legislations.

Salient Features of the New Seed Bill

Compulsory Registration of all kind and varieties being offered for sale by National Seed Board (NSB)

Under the Proposed Seeds Bill, 2004, all marketed seed and planting material, whether domestic or foreign, will have to be registered. This is a significant change from the existing law, which sought to regulate the quality of only a limited number of varieties notified under the law. Now, however, any seed for sowing or planting cannot be sold unless it is registered. All registered varieties will be record in a National Register of Seeds database. Registration will be granted for new varieties for a period of 15 years in the case of annuals and biennials crops and 18 years for long perennials. On expiry of the validated period the kind and variety of seed may be re-registered, for a like period by the Registration Sub Committee, based on agronomic performance of the kind and variety of seed under multi-location trials. Similarly the **Genetically Modified Seed or Transgenic Varieties** will be registered after conforming that the applicant has obtained clearance under the Environment (Protection) Act, 1986; A strict "No" to restriction technology and its products. The Registration Sub Committee is also empowered to cancel; or to revoke the registration certificate.

Compulsory Registration of imported seeds

All imported seed will also need to be registered, though the government may allow the import of unregistered seeds fro research purposes. Apart from the registration of imported seeds, the new bill does not make any other provisions, such as for phytosanitory standards, which rely on other existing legislations. However the main basis for the registration of imported seed is to support large companies importing seeds, which recently have been increasing substantially. For marketing seed lot that has been imported should meet the following requirements

- The provision of plants, fruits and seeds (regulation of import in India) order, 1989 or any corresponding order made under the section 3 of the destructive insects and pests acts, 1914.

- The minimum limits of germination, genetic and physical purity and seed health.

Compulsory Registration of seed producer, processor and dealer/ retailer

In India with an enactment of the 'Protection of Plant Variety and Farmers Right Act 2001; the breeder will get the royalty based on the commercial value of his variety. The information on the quantity of seed produced and sold will be important. Such information will also help in planning the seed programme of the country. To meet the requirement, the proposed seed bill 2004 has the provision of compulsory registration of seed producer to dealer.

Compulsory Registration of Horticultural Nurseries

Under the proposed seed bill, any propagating material is being considered as 'Seed". The provision for controlling the quality of vegetative propagating material such as cutting/ grafts had been made through compulsory registration of horticultural nurseries.

Seed Analysis and Seed Testing

Alike the Seed Acts (1966), under the proposed seed bill, the central government may, by notification, establish or declare any seed testing laboratory as the Central Seed Testing Laboratory (CSTL) to carry out the functions and trusted to the CSTL. The state government may, in consultation with the committee, by notification, establish one or more State Seed Testing laboratories in the government or non-government sector as a State Seed Testing Laboratory, were analysis of seed of any kind or variety shall be carried out. In order to accomplish these provisions, the state government may, by, notification; appoint such persons as it thinks fit, having the prescribed qualifications, to the seed inspectors or the seed analysts. Seeds are tested to confirm, that the seed lots offered for certification and certified seeds being sold in the market to meet out the minimum limits of germination, genetic purity and physical purity.

Seed Certification

Only the central and state seed testing; laboratories could offer seed certification under the old act; accredited individuals or institutions will be allowed under the new bill Seed Certification is mandatory and it allows self-certification. If also introduces a national register of sees and seeks to centralized registration by abrogating the authority of states to approve regional varieties.

Labeling requirement

Under the proposed seeds bill (2004), any seed that will be offered for sale in the market should meet out the following requirements:

- The variety /kind of that particular seed should be registered in the "National Seed Register".
- It should have a label to indicate that such seed conforms to the minimum limits of germination, genetic and physical purity, and seed health.
- Label should carry the information or other particulars, such as expected performance of the seed based on the information provided by the producer.

Farmer's interests and Seed Bill (proposed)

- In India over 70% of the farmers use their own saved seed and most of them are having marginal land holdings. Keeping these points in view, to safeguard the farmer's interest, the following provisions have been made in the proposed seed bill.
- Farmers who are involved in saving, using, exchanging, sharing or selling are exempted from compulsory registration, which otherwise, is a must for any person involved in seed trade.
- Farmers are not allowed to sell that are not meeting the minimum limits of germination, genetic and physical purity, and seed health.
- Under the proposed seeds bill, where the seed of any registered kind or variety is sold to a farmer, the producer, distributor or vender, as the case may be, shall disclose the expected performance of such kind or variety to the farmer under a given condition, and if, such registered seed fail to provide the expected performance under such condition, the farmer may claim compensation from the producer, distributor or vendor under the consumers Protection Act, 1986
- Similarly the research/scientific/educational/extension organizations will be exempted from all or any of the above provisions in the proposed Seed Bill, 2004.

The major omission in the seed bill is its lack of harmony with farmer's rights and plant breeder's rights on varieties provided under PPV FR. The seed bill will allowed unique monopolistic commercial advantage without acquiring plant breeder's right and without attracting the public interest liabilities attached to them.

Four most important public interest liabilities are as followed

- The liability for benefit sharing with tribal and farming communities the profit from seed trade, if their traditional varieties were used in the pedigree of the new seed
- The liability for compulsory licensing of variety on failure to supply seed commensurate with demand and at reasonable price
- The liability of truthfully declare the origin of the variety and its pedigree, which are important for benefit sharing
- The liability to give up exclusive marketing rights on completion of non-extendable 15 and 18 years for annual and perennial crop varieties respectively and the Seed bill nullify all these farmers friendly provisions of the PPV & FR.

Table contd.....

Table 1. Comparison of Seed Bill 2004 and Seed Act 1966*

	Seed Bill 2004	Seed Act 1966
Definition	'Agriculture' includes horticulture, forestry. Cultivation of plantation, medicinal and aromatic plants. Definitions of "Seed" and "Variety" have been changed to make them more specific and technical. Defines terms such as "Dealer", "Essentially Derived Variety", "Extant Variety", "Farmer", "Horticulture Nursery", "Misbranded", "Spurious Seed", and "Transgenic Variety".	' Agriculture' includes horticulture. Does not define these terms
Registration	All seeds for sale must be registered.	Only varieties notified by the government need to be registered
Seed Committee	Constitutes Central and State Seed Committees. A Registration Sub - Committee would register seeds of all varieties.	Constitutes Central Seed Committee. The central government, after consulting with the CSC, may notify a seed in order to regulate the quality of seed.
Transgenic Varieties	Special provi sions for registration of transgenic varieties of seeds.	No provision for transgenic varieties of seeds.
Compensation to Farmers	Provides for compensation to farmers under the Consumer Protection Act, 1986 in the event of under performance of seeds.	No specific provision for compensation mentioned in the Act.
Export and Import	All seed imports are regulated by the Plant Quarantine (Regulation of Import into India) Order, 2003 or any corresponding order of the Destructive Insects and Pests etc. 1914; shall conform to minimum limits of germination etc. Exports can be restricted if it adversely affects the food security of the country.	A person is restricted from exporting or importing notified variety of seed unless it conforms to minimum limits of germination etc.
Penalties	Any person who contravenes any provisions of the Act or imports, Sells or stocks seeds deemed to be misbranded or not registered can be punishable by a fine of Rs 5,000 to Rs 25,000. The penalty for giving false information is a imprison ment up to six months and/ or a fine up to Rs 50,0000.	Any person who contravenes any provisions of the Act. Prevents a Seed Inspector from taking samples etc. shall be punished for the first offence with a fine, which may extend to Rs 500. If the offence is repeated he may be imprisoned for a maximum term of six months and/or fined up to Rs 1,000.

Table 2. Comparison of Seeds Bill, 2004 and PPV & Act, 2001*

	Seed Bill 2004	**PPV FR & Act 2001**
Definitions	"Farmer" means any person who cultivates crops either by cultivation the land himself or through any other person but does not include any individual, company, trader or dealer who engages in the procurement and sale of seeds on a commercial basis.	"Farmer" means any person who cultivates crops by cultivating the land himself or cultivates crops by directly supervising the cultivation or land through any other person; or conserves and preserves, severally or jointly, with any other person any wild species or traditional varieties or adds value to such wild species or traditional varieties through selection and identification of their useful properties.
Registration	Establishes a Registration Sub-Committee, which would maintain a National Register of Seeds. No specification regarding parentage of variety. Registration is for 15 year for annual/biennial crops and 18 year for long duration perennials. On expiry, registration can be renewed for a similar period.	Establishes a Plant Varieties Registry, which would maintain a National Register of Plant Varieties. Specifies details under which a variety may be registered such as a complete passport data of the parental lines from which a variety has been derived. Registration is for 15 year for annual/biennial crops and 18 years for long duration perennials. Registration cannot be renewed.
Farmers' Rights	A farmer can save, use, exchange, share or sell his farm seeds and planting material. He cannot sell seeds under a brand name. Seeds sold have to conform to the minimum limit of germination, physical purity, genetic purity prescribed by the Act.	A farmer is entitled to save, use, sow, resow, exchange, share or sell his farm produce including seed of a variety protected under the Act in the same manner before this Act came into force. He cannot sell branded seed of a variety protected under the Act.
Compensation	The seed producer, distributor or vendor will have to disclose the expected performance of a particular variety of seed under certain given conditions. If the seed fails to perform to expected standards, the farmer can claim compensation from the dealer, distributor or vendor under the Consumer Protection Act, 1986.	If a breeder of a propagation material of a variety registered under the Act sells his product to a farmer, he has to disclose the expected performance under given conditions. If the propagation material fails to perform, the farmer can claim compensation in the prescribed manner before the Protection of Plant Varieties and Farmers' Rights Authority.

Contd...

Contd...

Penalties	Any person who contravenes any provisions of the Act, prevents a Seed Inspector from taking samples etc, shall be punished for the first offence with a fine up to Rs 500. If the offence is repeated he may be imprisoned up to six months and/or fined up to Rs 1,000	Penalty for applying false denomination to a variety is imprisonment up to two years and/or a fine between rupees 50,000 and rupees five lakh. Penalty for falsely representing a variety as registered is imprisonment up to three years and/or a fine between rupees one lakh and rupees five lakh or both. Penalty for subsequent offence is imprisonment up to three years and/or a fine between rupees two lakh and rupees twenty lakh.

•••

Annexure - I

International Treaty on Plant Genetic Resources for Food and Agriculture

PART I - INTRODUCTION

Article 1 - Objectives

1.1 The objectives of this Treaty are the conservation and sustainable use of plant genetic resources for food and agriculture and the fair and equitable sharing of the benefits arising out of their use, in harmony with the Convention on Biological Diversity, for sustainable agriculture and food security.

1.2 These objectives will be attained by closely linking this Treaty to the Food and Agriculture Organization of the United Nations and to the Convention on Biological Diversity.

Article 2 - Use of terms

For the purpose of this Treaty, the following terms shall have the meanings hereunder assigned to them. These definitions are not intended to cover trade in commodities: "*In situ* conservation" means the conservation of ecosystems and natural habitats and the maintenance and recovery of viable populations of species in their natural surroundings and, in the case of domesticated or cultivated plant species, in the surroundings where they have developed their distinctive properties.

"*Ex situ* conservation" means the conservation of plant genetic resources for food and agriculture outside their natural habitat.

"Plant genetic resources for food and agriculture" means any genetic material of plant origin of actual or potential value for food and agriculture.

"Genetic material" means any material of plant origin, including reproductive and vegetative propagating material, containing functional units of heredity.

"Variety" means a plant grouping, within a single botanical taxon of the lowest known rank, defined by the reproducible expression of its distinguishing and other genetic characteristics.

"*Ex situ* collection" means a collection of plant genetic resources for food and agriculture maintained outside their natural habitat.

"Centre of origin" means a geographical area where a plant species, either domesticated or wild,first developed its distinctive properties.

"Centre of crop diversity" means a geographic area containing a high level of genetic diversity for crop species in *in situ* conditions.

Article 3 - Scope

This Treaty relates to plant genetic resources for food and agriculture.

PART II - GENERAL PROVISIONS

Article 4 - General Obligations

Each Contracting Party shall ensure the conformity of its laws, regulations and procedures. With its obligations as provided in this Treaty.

Article 5 - Conservation, Exploration, Collection, Characterization, Evaluation and Documentation of Plant Genetic Resources for Food and Agriculture

5.1 Each Contracting Party shall, subject to national legislation, and in cooperation with other Contracting Parties where appropriate, promote an integrated approach to the exploration, conservation and sustainable use of plant genetic resources for food and agriculture and shall in particular, as appropriate:

(a) Survey and inventory plant genetic resources for food and agriculture, taking into account the status and degree of variation in existing populations, including those that are of potential use and, as feasible, assess any threats to them;

(b) Promote the collection of plant genetic resources for food and agriculture and relevant associated information on those plant genetic resources that are under threat or are of potential use;

(c) Promote or support, as appropriate, farmers and local communities' efforts to manage and conserve on-farm their plant genetic resources for food and agriculture;

(d) Promote *in situ* conservation of wild crop relatives and wild plants for food production, including in protected areas, by supporting, *inter alia*, the efforts of indigenous and local Communities;

(e) Cooperate to promote the development of an efficient and sustainable system of *ex situ* Conservation, giving due attention to the need for adequate documentation, characterization, regeneration and evaluation, and promote the development and transfer of appropriate technologies for this purpose with a view to improving the sustainable use of plant genetic resources for food and agriculture;

(f) Monitor the maintenance of the viability, degree of variation, and the genetic integrity of collections of plant genetic resources for food and agriculture.

5.2 The Contracting Parties shall, as appropriate, take steps to minimize or, if possible, eliminate threats to plant genetic resources for food and agriculture.

Article 6 - Sustainable Use of Plant Genetic Resources

6.1 The Contracting Parties shall develop and maintain appropriate policy and legal measures that promote the sustainable use of plant genetic resources for food and agriculture.

6.2 The sustainable use of plant genetic resources for food and agriculture may include such measures as:

(a) Pursuing fair agricultural policies that promote, as appropriate, the development and maintenance of diverse farming systems that enhance the sustainable use of agricultural biological diversity and other natural resources;

(b) strengthening research which enhances and conserves biological diversity by maximizing intra- and inter-specific variation for the benefit of farmers, especially those who generate and use their own varieties and apply ecological principles in maintaining soil fertility and in combating diseases, weeds and pests;

(c) promoting, as appropriate, plant breeding efforts which, with the participation of farmers, particularly in developing countries, strengthen the capacity to develop varieties particularly adapted to social, economic and ecological conditions, including in marginal areas;

(d) broadening the genetic base of crops and increasing the range of genetic diversity available to farmers;

(e) promoting, as appropriate, the expanded use of local and locally adapted crops, varieties and underutilized species;

(f) supporting, as appropriate, the wider use of diversity of varieties and species in on-farm management, conservation and sustainable use of crops and creating strong links to plant breeding and agricultural development in order to reduce crop vulnerability and genetic erosion, and promote increased world food production compatible with sustainable development; and

(g) reviewing, and, as appropriate, adjusting breeding strategies and regulations concerning variety release and seed distribution.

Article 7 - National Commitments and International Cooperation

7.1 Each Contracting Party shall, as appropriate, integrate into its agriculture and rural development policies and programmes, activities referred to in Articles 5 and 6, and cooperate with other Contracting Parties, directly or through FAO and other relevant international organizations, in the conservation and sustainable use of plant genetic resources for food and agriculture.

7.2 International cooperation shall, in particular, be directed to:

(a) establishing or strengthening the capabilities of developing countries and countries with economies in transition with respect to conservation and sustainable use of plant genetic resources for food and agriculture;

(b) enhancing international activities to promote conservation, evaluation, documentation, genetic enhancement, plant breeding, seed multiplication; and sharing, providing access to, and exchanging, in conformity with Part IV, plant genetic resources for food and agriculture and appropriate information and technology;

(c) maintaining and strengthening the institutional arrangements provided for in Part V; and

(d) implement the funding strategy of Article 18.

Article 8 - Technical Assistance

The Contracting Parties agree to promote the provision of technical assistance to Contracting

Parties, especially those that are developing countries or countries with economies in transition, either bilaterally or through the appropriate international organizations, with the objective of facilitating the implementation of this Treaty.

PART III - FARMERS' RIGHTS

Article 9 – Farmers' Rights

9.1 The Contracting Parties recognize the enormous contribution that the local and indigenous communities and farmers of all regions of the world, particularly those in the centres of origin and crop diversity, have made and will continue to make for the conservation and development of plant genetic resources which constitute the basis of food and agriculture production throughout the world.

9.2 The Contracting Parties agree that the responsibility for realizing Farmers' Rights, as they relate to plant genetic resources for food and agriculture, rests with national governments. In accordance with their needs and priorities, each Contracting Party should, as appropriate, and subject to its national legislation, take measures to protect and promote Farmers' Rights, including:

(a) protection of traditional knowledge relevant to plant genetic resources for food and agriculture;

(b) the right to equitably participate in sharing benefits arising from the utilization of plant genetic resources for food and agriculture; and

(c) the right to participate in making decisions, at the national level, on matters related to the conservation and sustainable use of plant genetic resources for food and agriculture.

9.3 Nothing in this Article shall be interpreted to limit any rights that farmers have to save, use, exchange and sell farm-saved seed/propagating material, subject to national law and as appropriate.

PART IV - THE MULTILATERAL SYSTEM OF ACCESS AND BENEFIT-SHARING

Article 10 – Multilateral System of Access and Benefit-sharing

10.1 In their relationships with other States, the Contracting Parties recognize the sovereign rights of States over their own plant genetic resources for food and agriculture, including that the authority to determine access to those resources rests with national governments and is subject to national legislation.

10.2 In the exercise of their sovereign rights, the Contracting Parties agree to establish a multilateral system, which is efficient, effective, and transparent, both to facilitate access to plant genetic resources for food and agriculture, and to share, in a fair and equitable way, the benefits arising from the utilization of these resources, on a complementary and mutually reinforcing basis.

Article 11 - Coverage of the Multilateral System

11.1 In furtherance of the objectives of conservation and sustainable use of plant genetic resources for food and agriculture and the fair and equitable sharing of benefits arising out of their use, as stated in Article 1, the Multilateral System shall cover the plant genetic resources for food and agriculture listed in Annex I, established according to criteria of food security and interdependence.

11.2 The Multilateral System, as identified in Article 11.1, shall include all plant genetic resources for food and agriculture listed in Annex I that are under the management and control of the Contracting Parties and in the public domain. With a view to achieving the fullest possible coverage of the Multilateral System, the Contracting Parties invite all other holders of the plant genetic resources for food and agriculture listed in Annex I to include these plant genetic resources for food and agriculture in the Multilateral System.

11.3 Contracting Parties also agree to take appropriate measures to encourage natural and legal persons within their jurisdiction who hold plant genetic resources for food and agriculture listed in Annex I to include such plant genetic resources for food and agriculture in the Multilateral System.

11.4 Within two years of the entry into force of the Treaty, the Governing Body shall assess the progress in including the plant genetic resources for food and agriculture referred to in paragraph 11.3 in the Multilateral System. Following this assessment, the Governing Body shall decide whether access shall continue to be facilitated to those natural and legal persons referred to in paragraph 11.3 that have not included these plant genetic resources for food and agriculture in the Multilateral System, or take such other measures as it deems appropriate.

11.5 The Multilateral System shall also include the plant genetic resources for food and agriculture listed in Annex I and held in the *ex situ* collections of the International Agricultural Research Centres of the Consultative Group on International Agricultural Research (CGIAR), as provided in Article 15.1a, and in other international institutions, in accordance with Article 15.5.

Article 12 - Facilitated access to plant genetic resources for food and agriculture within the Multilateral System

12.1 The Contracting Parties agree that facilitated access to plant genetic resources for food and agriculture under the Multilateral System, as defined in Article 11, shall be in accordance with the provisions of this Treaty.

12.2 The Contracting Parties agree to take the necessary legal or other appropriate measures to provide such access to other Contracting Parties through the Multilateral System. To this effect, such access shall also be provided to legal and natural persons under the jurisdiction of any Contracting Party, subject to the provisions of Article 11.4.

12.3 Such access shall be provided in accordance with the conditions below:

(a) Access shall be provided solely for the purpose of utilization and conservation for research, breeding and training for food and agriculture, provided that such purpose does not include chemical, pharmaceutical and/or other non-food/feed industrial uses. In the case of multiple-use crops (food and non-food), their importance for food security should be the determinant for their inclusion in the Multilateral System and availability for facilitated access.

(b) Access shall be accorded expeditiously, without the need to track individual accessions and free of charge, or, when a fee is charged, it shall not exceed the minimal cost involved;

(c) All available passport data and, subject to applicable law, any other associated available non-confidential descriptive information, shall be made available with the plant genetic resources for food and agriculture provided;

(d) Recipients shall not claim any intellectual property or other rights that limit the facilitated access to the plant genetic resources for food and agriculture, or their genetic parts or components, in the form received from the Multilateral System;

(e) Access to plant genetic resources for food and agriculture under development, including material being developed by farmers, shall be at the discretion of its developer, during the period of its development;

(f) Access to plant genetic resources for food and agriculture protected by intellectual and other property rights shall be consistent with relevant international agreements, and with relevant national laws;

(g) Plant genetic resources for food and agriculture accessed under the Multilateral System and conserved shall continue to be made available to the Multilateral System by the recipients of those plant genetic resources for food and agriculture, under the terms of this Treaty; and

(h) Without prejudice to the other provisions under this Article, the Contracting Parties agree that access to plant genetic resources for food and agriculture found in *in situ* conditions will be provided according to national legislation or, in the absence of such legislation, in accordance with such standards as may be set by the Governing Body.

12.4 To this effect, facilitated access, in accordance with Articles 12.2 and 12.3 above, shall be provided pursuant to a standard material transfer agreement (MTA), which shall be adopted by the Governing Body and contain the provisions of Articles 12.3a, d and g, as well as the benefitsharing provisions set forth in Article 13.2d(ii) and other relevant provisions of this Treaty, and the provision that the recipient of the plant genetic resources for food and agriculture shall require that the conditions of the MTA shall apply to the transfer of plant genetic resources for food and agriculture to another person or entity, as well as to any subsequent transfers of those plant genetic resources for food and agriculture.

12.5 Contracting Parties shall ensure that an opportunity to seek recourse is available,consistent with applicable jurisdictional requirements, under their legal systems, in case of contractual disputes arising under such MTAs, recognizing that obligations arising under such MTAs rest exclusively with the parties to those MTAs.

12.6 In emergency disaster situations, the Contracting Parties agree to provide facilitated access to appropriate plant genetic resources for food and agriculture in the Multilateral System for the purpose of contributing to the re-establishment of agricultural systems, in cooperation with disaster relief co-ordinators.

Article 13 - Benefit-sharing in the Multilateral System

13.1 The Contracting Parties recognize that facilitated access to plant genetic resources for food and agriculture which are included in the Multilateral System constitutes itself a major benefit of the Multilateral System and agree that benefits accruing therefrom shall be shared fairly and equitably in accordance with the provisions of this Article.

13.2 The Contracting Parties agree that benefits arising from the use, including commercial, of plant genetic resources for food and agriculture under the Multilateral System shall be shared fairly and equitably through the following mechanisms: the exchange of information, access to and transfer of technology, capacity-building, and the sharing of the benefits arising from commercialization, taking into account the priority activity areas in the rolling Global Plan of Action, under the guidance of the Governing Body:

(a) Exchange of information

The Contracting Parties agree to make available information which shall, *inter alia* ,encompass catalogues and inventories, information on technologies, results of technical,scientific and socio-economic research, including characterization, evaluation and utilization, regarding those plant genetic resources for food and agriculture under theMultilateral System. Such information shall be made available, where non-confidential,subject to applicable law and in accordance with national capabilities. Such informationshall be made available to all Contracting

Parties to this Treaty through the informationsystem, provided for in Article 17.

(b) Access to and transfer of technology

(i) The Contracting Parties undertake to provide and/or facilitate access totechnologies for the conservation, characterization, evaluation and use of plantgenetic resources for food and agriculture which are under the Multilateral System. Recognizing that some technologies can only be transferred throughgenetic material, the Contracting Parties shall provide and/or facilitate access tosuch technologies and genetic material which is under the Multilateral Systemand to improved varieties and genetic material developed through the use of plantgenetic resources for food and agriculture under the Multilateral System, inconformity with the provisions of Article 12. Access to these technologies,improvedvarieties and genetic material shall be provided and/or facilitated, whilerespecting applicable property rights and access laws, and in accordance with national capabilities.

(ii) Access to and transfer of technology to countries, especially to developingcountries and countries with economies in transition, shall be carried out througha set of measures, such as the establishment and maintenance of, and participation in, crop-based thematic groups on utilization of plant geneticresources for food and agriculture, all types of partnership in research anddevelopment and in commercial joint ventures relating to the material received, human resource development, and effective access to research facilities.

(iii) Access to and transfer of technology as referred to in (i) and (ii) above, including that protected by intellectual property rights, to developing countries that are Contracting Parties, in particular least developed countries, and countries with economies in transition, shall be provided and/or facilitated under fair and most favourable terms, in particular in the case of technologies for use in conservation as well as technologies for the benefit of farmers in developing countries, especially in least developed countries, and countries with economies in transition, including on concessional and preferential terms where mutually agreed, *inter alia*, through partnerships in research and development under the Multilateral System. Such access and transfer shall be provided on terms which recognize and are consistent with the adequate and effective protection of intellectual property rights.

(c) *Capacity-building*

Taking into account the needs of developing countries and countries with economies in transition, as expressed through the priority they accord to building capacity in plant genetic resources for food and agriculture in their plans and programmes, when in place,in respect of those plant genetic resources for food and agriculture covered by the Multilateral System, the Contracting Parties agree to give priority to (i) establishing and/or strengthening programmes for scientific and technical education and training in conservation and sustainable use of plant genetic resources for food and agriculture, (ii) developing and strengthening facilities for conservation and sustainable use of plant genetic resources for food and agriculture, in particular in developing countries, and countries with economies in transition, and (iii) carrying out scientific research preferably, and where possible, in developing countries and countries with economies in transition, in cooperation with institutions of such countries, and developing capacity for such research in fields where they are needed.

(d) *Sharing of monetary and other benefits of commercialization*

(i) The Contracting Parties agree, under the Multilateral System, to take measures in order to achieve commercial benefit-sharing, through the involvement of the private and public sectors in activities identified under this Article, through partnerships and collaboration, including with the private sector in developing countries and countries with economies in transition, in research and technology development;

(ii) The Contracting Parties agree that the standard Material Transfer Agreement referred to in Article 12.4 shall include a requirement that a recipient who commercializes a product that is a plant genetic resource for food and agriculture and that incorporates material accessed from the Multilateral System, shall pay to the mechanism referred to in Article 19.3f, an equitable share of the benefits arising from the commercialization of that product, except whenever such a product is available without restriction to others for further research and breeding, in which case the recipient who commercializes shall be encouraged to make such payment.

The Governing Body shall, at its first meeting, determine the level, form and manner of the payment, in line with commercial practice. The Governing Body may decide to establish different levels of payment for various categories of recipients who commercialize such products; it may also decide on the need to exempt from such payments small farmers in developing countries and in countries with economies in transition. The Governing Body may, from time to time, review the levels of payment with a view to achieving fair and equitable sharing of benefits, and it may also assess, within a period of five years from the entry into force of this Treaty, whether the mandatory payment requirement in the MTA shall apply also in cases where such commercialized products are available without restriction to others for further research and breeding.

13.3 The Contracting Parties agree that benefits arising from the use of plant genetic resources for food and agriculture that are shared under the Multilateral System should flow primarily, directly and indirectly, to farmers in all countries, especially in developing countries, and countries with economies in transition, who conserve and sustainably utilize plant genetic resources for food and agriculture.

13.4 The Governing Body shall, at its first meeting, consider relevant policy and criteria for specific assistance under the agreed funding strategy established under Article 18 for the conservation of plant genetic resources for food and agriculture in developing countries, and countries with economies in transition whose contribution to the diversity of plant genetic resources for food and agriculture in the Multilateral System is significant and/or which have special needs.

13.5 The Contracting Parties recognize that the ability to fully implement the Global Plan of Action, in particular of developing countries and countries with economies in transition, will depend largely upon the effective implementation of this Article and of the funding strategy as provided in Article 18.

13.6 The Contracting Parties shall consider modalities of a strategy of voluntary benefit sharing contributions whereby Food Processing Industries that benefit from plant genetic resources for food and agriculture shall contribute to the Multilateral System.

PART V - SUPPORTING COMPONENTS

Article 14 – Global Plan of Action

Recognizing that the rolling Global Plan of Action for the Conservation and Sustainable

Use of Plant Genetic Resources for Food and Agriculture is important to this Treaty, Contracting Parties should promote its effective implementation, including through national actions and, as appropriate, international cooperation to provide a coherent framework, *inter alia,* for capacity building, technology transfer and exchange of information, taking into account the provisions of Article 13.

Article 15 - Ex Situ Collections of Plant Genetic Resources for Food and Agriculture held by the International Agricultural Research Centres of the Consultative Group on International Agricultural Research and other International Institutions

15.1 The Contracting Parties recognize the importance to this Treaty of the *ex situ* collections of plant genetic resources for food and agriculture held in trust by the International Agricultural Research Centres (IARCs) of the Consultative Group on International Agricultural Research (CGIAR). The Contracting Parties call upon the IARCs to sign agreements with the Governing Body with regard to such *ex situ* collections, in accordance with the following terms and conditions:

(a) Plant genetic resources for food and agriculture listed in Annex I of this Treaty and held by the IARCs shall be made available in accordance with the provisions set out in Part IV of this Treaty.

(b) Plant genetic resources for food and agriculture other than those listed in Annex I of this Treaty and collected before its entry into force that are held by IARCs shall be made available in accordance with the provisions of the MTA currently in use pursuant to agreements between the IARCs and the FAO. This MTA shall be amended by the Governing Body no later than its second regular session, in consultation with the IARCs, in accordance with the relevant provisions of this Treaty, especially Articles 12 and 13, and under the following conditions:

(i) The IARCs shall periodically inform the Governing Body about the MTAs entered into, according to a schedule to be established by the Governing Body;

(ii) The Contracting Parties in whose territory the plant genetic resources for food and agriculture were collected from *in situ* conditions shall be provided with samples of such plant genetic resources for food and agriculture on demand, without any MTA;

(iii) Benefits arising under the above MTA that accrue to the mechanism mentioned in Article 19.3f shall be applied, in particular, to the conservation and sustainable use of the plant genetic resources for food and agriculture in question, particularly in national and regional programmes in developing countries and countries with economies in transition, especially in centres of diversity and the least developed countries; and

(iv) The IARCs shall take appropriate measures, in accordance with their capacity, to maintain effective compliance with the conditions of the MTAs, and shall promptly inform the Governing Body of cases of non-compliance.

(c) IARCs recognize the authority of the Governing Body to provide policy guidance relating to *ex situ* collections held by them and subject to the provisions of this Treaty.

(d) The scientific and technical facilities in which such *ex situ* collections are conserved shall remain under the authority of the IARCs, which undertake to manage and administer these *ex situ* collections in accordance with internationally accepted standards, in particular the Genebank Standards as endorsed by the FAO Commission on Genetic Resources for Food and Agriculture.

(e) Upon request by an IARC, the Secretary shall endeavour to provide appropriate technical support.

(f) The Secretary shall have, at any time, right of access to the facilities, as well as right to inspect all activities performed therein directly related to the conservation and exchange of the material covered by this Article.

(g) If the orderly maintenance of these *ex situ* collections held by IARCs is impeded or threatened by whatever event, including *force majeure*, the Secretary, with the approval of the host country, shall assist in its evacuation or transfer, to the extent possible.

15.2 The Contracting Parties agree to provide facilitated access to plant genetic resources for food and agriculture in Annex I under the Multilateral System to IARCs of the CGIAR that have signed agreements with the Governing Body in accordance with this Treaty. Such Centres shall be included in a list held by the Secretary to be made available to the Contracting Parties on request.

15.3 The material other than that listed in Annex I, which is received and conserved by IARCs after the coming into force of this Treaty, shall be available for access on terms consistent with those mutually agreed between the IARCs that receive the material and the country of origin of such resources or the country that has acquired those resources in accordance with the Convention on Biological Diversity or other applicable law.

15.4 The Contracting Parties are encouraged to provide IARCs that have signed agreements with the Governing Body with access, on mutually agreed terms, to plant genetic resources for food and agriculture not listed in Annex I that are important to the programmes and activities of the IARCs.

15.5 The Governing Body will also seek to establish agreements for the purposes stated in this Article with other relevant international institutions.

Article 16 – International Plant Genetic Resources Networks

16.1 Existing cooperation in international plant genetic resources for food and agriculture networks will be encouraged or developed on the basis of existing arrangements and consistent with the terms of this Treaty, so as to achieve as complete coverage as possible of plant genetic resources for food and agriculture.

16.2 The Contracting Parties will encourage, as appropriate, all relevant institutions, including governmental, private, non-governmental, research, breeding and other institutions, to participate in the international networks.

Article 17 – The Global Information System on Plant Genetic Resources for Food and Agriculture

17.1 The Contracting Parties shall cooperate to develop and strengthen a global information system to facilitate the exchange of information, based on existing information systems, onscientific, technical and environmental matters related to plant genetic resources for food and agriculture, with the expectation that such exchange of information will contribute to the sharing of benefits by making information on plant genetic resources for food and agriculture available to all Contracting Parties. In developing the Global Information System, cooperation will be sought with the Clearing House Mechanism of the Convention on Biological Diversity.

17.2 Based on notification by the Contracting Parties, early warning should be provided about hazards that threaten the efficient maintenance of plant genetic resources for food and agriculture, with a view to safeguarding the material.

17.3 The Contracting Parties shall cooperate with the Commission on Genetic Resources for Food and Agriculture of the FAO in its periodic reassessment of the state of the world's plant genetic resources for food and agriculture in order to facilitate the updating of the rolling Global Plan of Action referred to in Article 14.

PART VI - FINANCIAL PROVISIONS

Article 18 – Financial Resources

18.1 The Contracting Parties undertake to implement a funding strategy for the implementation of this Treaty in accordance with the provisions of this Article.

18.2 The objectives of the funding strategy shall be to enhance the availability, transparency, efficiency and effectiveness of the provision of financial resources to implement activities under this Treaty.

18.3 In order to mobilize funding for priority activities, plans and programmes, in particular in developing countries and countries with economies in transition, and taking the Global Plan of Action into account, the Governing Body shall periodically establish a target for such funding.

18.4 Pursuant to this funding strategy:

(a) The Contracting Parties shall take the necessary and appropriate measures within the Governing Bodies of relevant international mechanisms, funds and bodies to ensure due priority and attention to the effective allocation of predictable and agreed resources for the implementation of plans and programmes under this Treaty.

(b) The extent to which Contracting Parties that are developin g countries and Contracting Parties with economies in transition will effectively implement their commitments under this Treaty will depend on the effective allocation, particularly by the developed country Parties, of the resources referred to in this Article. Contracting Parties that are developing countries and Contracting Parties with economies in transition will accord due priority in their own plans and programmes to building capacity in plant genetic resources for food and agriculture.

(c) The Contracting Parties that are developed countries also provide, and Contracting Parties that are developing countries and Contracting Parties with economies in transition avail themselves of, financial resources for the implementation of this Treaty through bilateral and regional and multilateral channels. Such channels shall include the mechanism referred to in Article 19.3f.

(d) Each Contracting Party agrees to undertake, and provide financial resources for national activities for the conservation and sustainable use of plant genetic resources for food and agriculture in accordance with its national capabilities and financial resources. The financial resources provided shall not be used to ends inconsistent with this Treaty, inparticular in areas related to international trade in commodities. ;

(e) The Contracting Parties agree that the financial benefits arising from Article 13.2d are part of the funding strategy.

(f) Voluntary contributions may also be provided by Contracting Parties, the private sector, taking into account the provisions of Article 13, non-governmental organisations and other sources. The Contracting Parties agree that the Governing Body shall consider modalities of a strategy to promote such contributions;

18.5 The Contracting Parties agree that priority will be given to the implementation of agreed plans and programmes for farmers in developing countries, especially in least developed countries, and in countries with economies in transition, who conserve and sustainably utilize plant genetic resources for food and agriculture.

PART VII - INSTITUTIONAL PROVISIONS

Article 19 – Governing Body

19.1 A Governing Body for this Treaty is hereby established, composed of all Contracting Parties.

19.2 All decisions of the Governing Body shall be taken by consensus unless by consensus another method of arriving at a decision on certain measures is reached, except that consensus shall always be required in relation to Articles 23 and 24.

19.3 The functions of the Governing Body shall be to promote the full implementation of this Treaty, keeping in view its objectives, and, in particular, to:

(a) provide policy direction and guidance to monitor, and adopt such recommendations as necessary for the implementation of this Treaty and, in particular, for the operation of the Multilateral System;

(b) adopt plans and programmes for the implementation of this Treaty;

(c) adopt, at its first session, and periodically review the funding strategy for the implementation of this Treaty, in accordance with the provisions of Article 18;

(d) adopt the budget of this Treaty;

(e) consider and establish subject to the availability of necessary funds such subsidiary bodies as may be necessary, and their respective mandates and composition;

(f) establish, as needed, an appropr iate mechanism, such as a Trust Account, for receiving and utilizing financial resources that will accrue to it for purposes of implementing this Treaty;

(g) establish and maintain cooperation with other relevant international organizations and treaty bodies, including in particular the Conference of the Parties to the Convention on Biological Diversity, on matters covered by this Treaty, including their participation in the funding strategy;

(h) consider and adopt, as required, amendments to this Treaty, in accordance with the provisions of Article 23;

(i) consider and adopt, as required, amendments to annexes to this Treaty, in accordance with the provisions of Article 24;

(j) consider modalities of a strategy to encourage voluntary contributions, in particular, with reference to Articles 13 and 18;

(k) perform such other functions as may be necessary for the fulfilment of the objectives of this Treaty;

(l) take note of relevant decisions of the Conference of the Parties to the Convention on Biological Diversity and other relevant international organizations and treaty bodies;

(m) inform, as appropriate, the Conference of the Parties to the Convention on Biological Diversity and other relevant international organizations and treaty bodies of matters regarding the implementation of this Treaty; and

(n) approve the terms of agreements with the IARCs and other international institutions under Article 15, and review and amend the MTA in Article 15.

19.4 Subject to Article 19.6, each Contracting Party shall have one vote and may be represented at sessions of the Governing Body by a single delegate who may be accompanied by an alternate, and by experts and advisers. Alternates, experts and advisers may take part in the proceedings of the Governing Body but may not vote, except in the case of their being duly authorized to substitute for the delegate.

19.5 The United Nations, its specialized agencies and the International Atomic Energy Agency, as well as any State not a Contracting Party to this Treaty, may be represented as observers at sessions of the Governing Body. Any other body or agency, whether governmental or non-governmental, qualified in fields relating to conservation and sustainable use of plant genetic resources for food and agriculture, which has informed the Secretary of its wish to be represented as an observer at a session of the Governing Body, may be admitted unless at least one third of the Contracting Parties present object. The admission and participation of observers shall be subject to the Rules of Procedure adopted by the Governing Body.

19.6 A Member Organization of FAO that is a Contracting Party and the member states of that Member Organization that are Contracting Parties shall exercise their membership rights and fulfil their membership obligations in accordance, *mutatis mutandis*, with the Constitution and General Rules of FAO.

19.7 The Governing Body shall adopt and amend, as required, its own Rules of Procedure and financial rules which shall not be inconsistent with this Treaty.

19.8 The presence of delegates representing a majority of the Contracting Parties shall be necessary to constitute a quorum at any session of the Governing Body.

19.9 The Governing Body shall hold regular sessions at least once every two years. These sessions should, as far as possible, be held back-to-back with the regular sessions of the Commission on Genetic Resources for Food and Agriculture.

19.10 Special Sessions of the Governing Body shall be held at such other times as may be deemed necessary by the Governing Body, or at the written request of any Contracting Party, provided that this request is supported by at least one third of the Contracting Parties.

19.11 The Governing Body shall elect its Chairperson and Vice-Chairpersons (collectively referred to as "the Bureau"), in conformity with its Rules of Procedure.

Article 20 - Secretary

20.1 The Secretary of the Governing Body shall be appointed by the Director-General of FAO, with the approval of the Governing Body. The Secretary shall be assisted by such staff as may be required.

20.2 The Secretary shall perform the following functions:

(a) arrange for and provide administrative support for sessions of the Governing Body and for any subsidiary bodies as may be established;

(b) assist the Governing Body in carrying out its functions, including the performance of specific tasks that the Governing Body may decide to assign to it;

(c) report on its activities to the Governing Body.

20.3 The Secretary shall communicate to all Contracting Parties and to the Director-General:

(a) decisions of the Governing Body within sixty days of adoption;

(b) information received from Contracting Parties in accordance with the provisions of this Treaty.

20.4 The Secretary shall provide documentation in the six languages of the United Nations for sessions of the Governing Body.

20.5 The Secretary shall cooperate with other organizations and treaty bodies, including in particular the Secretariat of the Convention on Biological Diversity, in achieving the objectives of this Treaty.

Article 21 - Compliance

The Governing Body shall, at its first meeting, consider and approve cooperative and effective procedures and operational mechanisms to promote compliance with the provisions of this Treaty and to address issues of non-compliance. These procedures and mechanisms shall minclude monitoring, and offering advice or assistance, including legal advice or legal assistance, when needed, in particular to developing countries and countries with economies in transition.

Article 22 - Settlement of Disputes

22.1 In the event of a dispute between Contracting Parties concerning the interpretation or application of this Treaty, the parties concerned shall seek solutions by negotiation.

22.2 If the parties concerned cannot reach agreement by ne gotiation, they may jointly seek the good offices of, or request mediation by, a third party.

22.3 When ratifying, accepting, approving or acceding to this Treaty, or at any time thereafter, a Contracting Party may declare in writing to the Depositary that for a dispute not resolved in accordance with Article 22.1 or Article 22.2 above, it accepts one or both of the following means of dispute settlement as compulsory:

(a) Arbitration in accordance with the procedure laid down in Part 1 of Annex II to this Treaty;

(b) Submission of the dispute to the International Court of Justice. 22.4 If the parties to the dispute have not, in accordance with Article 22.3 above, accepted the same or any procedure, the dispute shall be submitted to conciliation in accordance with Part 2 of Annex II to this Treaty unless the parties otherwise agree.

Article 23 - Amendments of the Treaty

23.1 Amendments to this Treaty may be proposed by any Contracting Party.

23.2 Amendments to this Treaty shall be adopted at a session of the Governing Body. The text of any proposed amendment shall be communicated to Contracting Parties by the Secretary at least six months before the session at which it is proposed for adoption.

23.3 All amendments to this Treaty shall only be made by consensus of the Contracting Parties present at the session of the Governing Body.

23.4 Any amendment adopted by the Governing Body shall come into force among Contracting Parties having ratified, accepted or approved it on the ninetieth day after the deposit of instruments of ratification, acceptance or approval by two-thirds of the Contracting Parties. Thereafter the amendment shall enter into force for any other Contracting Party on the ninetieth day after that Contracting Party deposits its instrument of ratification, acceptance or approval of the amendment.

23.5 For the purpose of this Article, an instrument deposited by a Member Organization of FAO shall not be counted as additional to those deposited by member states of such an organization.

Article 24 - Annexes

24.1 The annexes to this Treaty shall form an integral part of this Treaty and a reference to this Treaty shall constitute at the same time a reference to any annexes thereto.

24.2 The provisions of Article 23 regarding amendments to this Treaty shall apply to the amendment of annexes.

Article 25 - Signature

This Treaty shall be open for signature at the FAO from 3 November 2001 to 4 November 2002 by all Members of FAO and any States that are not Members of FAO but are Members of the United Nations, or any of its specialized agencies or of the International Atomic Energy Agency.

Article 26 - Ratification, Acceptance or Approval

This Treaty shall be subject to ratification, acceptance or approval by the Members and non-Members of FAO referred to in Article 25. Instruments of ratification, acceptance, or approval shall be deposited with the Depositary.

Article 27 - Accession

This Treaty shall be open for accession by all Members of FAO and any States that are not Members of FAO but are Members of the United Nations, or any of its specialized agencies or of the International Atomic Energy Agency from the date on which the Treaty is closed for signature. Instruments of accession shall be deposited with the Depositary.

Article 28 - Entry into force

28.1 Subject to the provisions of Article 29.2, this Treaty shall enter into force on the ninetieth day after the deposit of the fortieth instrument of ratification, acceptance, approval or accession, provided that at least twenty instruments of ratification, acceptance, approval or accession have been deposited by Members of FAO.

28.2 For each Member of FAO and any State that is not a Member of FAO but is a Member of the United Nations, or any of its specialized agencies or of the International Atomic Energy Agency that ratifies, accepts, approves or accedes to this Treaty after the deposit, in accordance with Article 28.1, of the fortieth instrument of ratification, acceptance, approval or accession, the Treaty shall enter into force on the ninetieth day following the deposit of its instrument of ratification, acceptance, approval or accession.

Article 29 - Member Organizations of FAO

29.1 When a Member Organization of FAO deposits an instrument of ratification, acceptance, approval or accession for this Treaty, the Member Organization shall, in accordance with the provisions of Article II.7 of the FAO Constitution, notify any change regarding its distribution of competence to its declaration of competence submitted under Article II.5 of the FAO Constitution as may be necessary in light of its acceptance of this Treaty. Any Contracting Party to this Treaty may, at any time, request a Member Organization of FAO that is a Contracting Party to this Treaty to provide information as to which, as between the Member Organization and its member states, is responsible for the implementation of any particular matter covered by this Treaty. The Member Organization shall provide this information within a reasonable time.

29.2 Instruments of ratification, acceptance, approval, accession or withdrawal, deposited by a Member Organization of FAO, shall not be counted as additional to those deposited by its Member States.

Article 30 - Reservations

No reservations may be made to this Treaty.

Article 31 - Non-Parties

The Contracting Parties shall encourage any Member of FAO or other State, not a Contracting Party to this Treaty, to accept this Treaty.

Article 32 - Withdrawals

32.1 Any Contracting Party may at any time after two years from the date on which this Treaty has entered into force for it, notify the Depositary in writing of its withdrawal from this Treaty. The Depositary shall at once inform all Contracting Parties.

32.2 Withdrawal shall take effect one year from the date of receipt of the notification.

Article 33 - Termination

33.1 This Treaty shall be automatically terminated if and when, as the result of withdrawals,the number of Contracting Parties drops below forty, unless the remaining Contracting Parties unanimously decide otherwise.

33.2 The Depositary shall inform all remaining Contracting Parties when the number of Contracting Parties has dropped to forty.

33.3 In the event of termination the disposition of assets shall be governed by the financial rules to be adopted by the Governing Body.

Article 34 - Depositary

The Director-General of FAO shall be the Depositary of this Treaty.

Article 35 - Authentic Texts

The Arabic, Chinese, English, French, Russian and Spanish texts of this Treaty are Equally authentic.

ANNEX - II

PART 1 ARBITRATION

Article 1

The claimant party shall notify the Secretary that the parties to the dispute are referring it to arbitration pursuant to Article 22. The notification shall state the subject-matter of arbitration and include, in particular, the articles of this Treaty, the interpretation or application of which are at issue. If the parties to the dispute do not agree on the subject matter of the dispute before the President of the tribunal is designated, the arbitral tribunal shall determine the subject matter. The Secretary shall forward the information thus received to all Contracting Parties to this Treaty.

Article 2

1. In disputes between two parties to the dispute, the arbitral tribunal shall consist of three members. Each of the parties to the dispute shall appoint an arbitrator and the two arbitrators so appointed shall designate by common agreement the third arbitrator who shall be the President of the tribunal. The latter shall not be a national of one of the parties to the dispute, nor have his or her usual place of residence in the territory of one of these parties to the dispute, nor be employed by any of them, nor have dealt with the case in any other capacity.
2. In disputes between more than two Contracting Parties, parties to the dispute with the same interest shall appoint one arbitrator jointly by agreement.
3. Any vacancy shall be filled in the manner prescribed for the initial appointment.

Article 3

1. If the President of the arbitral tribunal has not been designated within two months of the appointment of the second arbitrator, the Director-General of FAO shall, at the request of a party to the dispute, designate the President within a further two-month period.
2. If one of the parties to the dispute does not appoint an arbitrator within two months of receipt of the request, the other party may

inform the Director-General of FAO who shall make the designation within a further two-month period.

Article 4

The arbitral tribunal shall render its decisions in accordance with the provisions of this Treaty and international law.

Article 5

Unless the parties to the dispute otherwise agree, the arbitral tribunal shall determine its own rules of procedure.

Article 6

The arbitral tribunal may, at the request of one of the parties to the dispute, recommend essential interim measures of protection.

Article 7

The parties to the dispute shall facilitate the work of the arbitral tribunal and, in particular, using all means at their disposal, shall:

(a) Provide it with all relevant documents, information and facilities; and

(b) Enable it, when necessary, to call witnesses or experts and receive their evidence.

Article 8

The parties to the dispute and the arbitrators are under an obligation to protect the confidentiality of any information they receive in confidence during the proceedings of the arbitral tribunal.

Article 9

Unless the arbitral tribunal determines otherwise because of the particular circumstances of the case, the costs of the tribunal shall be borne by the parties to the dispute in equal shares. The tribunal shall keep a record of all its costs, and shall furnish a final statement thereof to the parties to the dispute.

Article 10

Any Contracting Party that has an interest of a legal nature in the subject-matter of the dispute which may be affected by the decision in the case, may intervene in the proceedings with the consent of the tribunal.

Article 11

The tribunal may hear and determine counterclaims arising directly out of the subject-matter of the dispute.

Article 12

Decisions both on procedure and substance of the arbitral tribunal shall be taken by a majority vote of its members.

Article 13

If one of the parties to the dispute does not appear before the arbitral tribunal or fails to defend its case, the other party may request the tribunal to continue the proceedings and to make its award. Absence of a party to the dispute or a failure of a party to the dispute to defend its case shall not constitute a bar to the proceedings. Before rendering its final decision, the arbitral tribunal must satisfy itself that the claim is well founded in fact and law.

Article 14

The tribunal shall render its final decision within five months of the date on which it is fully constituted unless it finds it necessary to extend the time-limit for a period which should not exceed five more months.

Article 15

The final decision of the arbitral tribunal shall be confined to the subject-matter of the dispute and shall state the reasons on which it is based. It shall contain the names of the members who have participated and the date of the final decision. Any member of the tribunal may attach a separate or dissenting opinion to the final decision.

Article 16

The award shall be binding on the parties to the dispute. It shall be without appeal unless the parties to the dispute have agreed in advance to an appellate procedure.

Article 17

Any controversy which may arise between the parties to the dispute as regards the interpretation or manner of implementation of the final decision may be submitted by either party to the dispute for decision to the arbitral tribunal which rendered it.

PART 2 CONCILIATION

Article 1

A conciliation commission shall be created upon the request of one of the parties to the dispute. The commission shall, unless the parties to the dispute otherwise agree, be composed of five members, two appointed by each party concerned and a President chosen jointly by those members.

Article 2

In disputes between more than two Contracting Parties, parties to the dispute with the same interest shall appoint their members of the commission jointly by agreement. Where two or more parties to the dispute have separate interests or there is a disagreement as to whether they are of the same interest, they shall appoint their members separately.

Article 3

If any appointments by the parties to the dispute are not made within two months of the date of the request to create a conciliation commission, the Director-General of FAO shall, if asked to do so by the party to the dispute that made the request, make those appointments within a further two-month period.

Article 4

If a President of the conciliation commission has not been chosen within two months of the last of the members of the commission being appointed, the Director-General of FAO shall, if asked to do so by a party to the dispute, designate a President within a further two-month period.

Article 5

The conciliation commission shall take its decisions by majority vote of its members. It shall, unless the parties to the dispute otherwise agree, determine its own procedure. It shall render a proposal for resolution of the dispute, which the parties shall consider in good faith.

Article 6

A disagreement as to whether the conciliation commission has competence shall be decided by the commission.

A useful system is that of Harlan and de Wet (1971), which recognizes three different gene pools according to the difficulty in transfer between them.

The Primary gene pool (GP-1) corresponds to the traditional concept of a biological species; cultivated and wild races are included in subspecies A and subspecies B , respectively .The number of subspecies in A is generally one. Several authentic B subspecies can be defined, their relationships with the cultivated one (or ones) ranging from very close to rather far , but always within limit that they easily intercross with A to produce fertile hybrids cross between forms belonging to this gene pool must be considered intraspecific.

The secondary gene pool (GP-2) includes the biological species that will cross with the species under consideration. In Harlan and deWet,s words, "gene transfer is possible , but one must struggle with those barriers that separate biological species" .In spite of the high sterility level of the hybrids, poor (if any) chromosome pairing ,and poor development of hybrids, both interspecific crosses and gene transfer are possible without special techniques, such as embryo culture. The allocation of a species to the primary or the secondary gene pool can be rather difficult in some group.

The tertiary gene pool (GP-3) includes species that can cross with that under study, but with the assistance of special techniques e.g. embryo culture , via artificially produced tetraploids, grafting, tissue culture, and the use of bridging species.

•••

Annexure - II

Preparation of Buffers

1. Acetate buffer (0.2 M)

Stock solutions

A. 0.2M Acetic acid solution = Dilute 11.55 ml of acetic acid to 1000ml with water.

B. 0.2M Sodium acetate solution ($C_2 H_3 O_2$ Na. 3H2 O) = 22.22g in water and make up to 1000ml.

Mix x ml of A with y ml of B.

x ml	9.25	8.8	8.2	7.35	6.3	5.1	4.1	3.0	2.1	1.4	0.9	0.6
y ml	0.75	4.2	1.8	2.65	3.7	4.9	5.9	7.0	7.9	8.6	9.1	9.4
pH	3.6	3.8	4.0	4.2	4.4	4.6	4.8	5.0	5.2	5.4	5.6	5.8

2. Carbonate/bicarbonate buffer*

Stock solutions

A. 0.2M Solution of anhydrous sodium carbonate = 21.2g in water and make up to 1000ml

B. 0.2M Solution of sodium bicarbonate = 16.8g in water and make up to 1000ml.

x ml of A is mixed with y ml of B and diluted to 200ml with water.

x ml	4.0	7.5	9.5	13	16	19.5	22	25	27.5	30	33	35.5	38.5	40.5
y ml	46	42.5	40.5	37	34	30.5	28	25	22.5	20	17	14.5	11.5	9.5
pH	9.2	9.3	9.4	9.5	9.6	9.7	9.8	9.9	10.0	10.1	10.2	10.3	10.4	10.5

* Delory, G.E and J. King (1945). *Biochem. J.* **39**: 245

3. Glycine-HCI buffer*

Stock solutions

A. 0.2M Solution of glycine = 15.01g in water and make up to 1000ml.

B. 0.2M HCI

50ml of A is mixed with xml of B and diluted to total of 200ml with water.

x ml	5.0	6.4	8.2	11.2	16.8	24.2	32.4	44.0
pH	3.6	3.4	3.2	3.0	2.8	2.6	2.4	2.2

* Sorenson, S.P.L. (1909). *Biochem. Z.*, 21: 131; 22: 352.

4. Glycine-NaOH buffer*

Stock solutions

A. 0.2M Solution of glycine = 15.01g in water and make up to 1000ml.

B. 0.2M NaOH

50ml of A is mixed with xml of B and diluted to total of 200ml with water.

x ml	4.0	6.0	8.8	12.0	16.8	22.4	27.2	32	38.6	45.5
pH	8.6	8.8	9.0	9.2	9.4	9.6	9.8	10	10.4	10.6

* Sorenson, S.P.L. (1909). *Biochem. Z.*, 21: 131; 22: 352.

5. Hydrochloric acid – potassium chloride buffer*

Stock solutions

A. 0.2M Solution of KCI = 14.91g in water and make up to 1000ml.

B. 0.2M HCI

50ml of A is mixed with xml of B and diluted to total of 200ml with water.

x ml	97.0	78.0	64.5	51.0	41.5	33.3	26.3	20.6	16.6	13.2	10.6	8.4	6.7
pH	1.0	1.1	1.2	1.3	1.4	1.5	1.6	1.7	1.8	1.9	2.0	2.1	2.2

* Clark, W.M. and H.A. Lubs (1917). *J. Bacteriol.* 2 : 1

6. Sodium phosphate buffer (0.1M)*

Stock solutions

A. 0.2M Solution of $Na_2\ HPO_4 . 7H_2\ O$ = 35.61g in water and make up to 1000ml.

B. 0.2M HCI

50ml of A is mixed with xml of B an diluted to total of 200ml with water.

x ml	8.0	12.3	18.5	26.4	37.5	49	61	72	81	87	91.5	94.7
y ml	92	87.7	81.5	73.5	62.51	51	39	28	19	13	8.5	5.3
pH	5.8	6.0	6.2	6.4	6.6	6.8	7.0	7.2	7.4	7.6	7.8	8.0

* Sorenson, S.P.L. (1909). *Biochem. Z.*, 21: 131; 22: 352.

7. Potassium phosphate buffer (0.1M)

Stock solutions

A. 0.2M Solution of KOH = 11.22g in water and make up to 1000ml.

B. 0.2M Solution of $KH_2 PO_4$ = 27.22g in water and make up to 1000ml.

Mix x ml of A with y ml of B and dilute to 100ml with water.

x ml	3.5	5.8	9.1	13.0	18.0	24.0	30.0	35	40	43	45	47
y ml	50	50	50	50	50	50	50	50	50	50	50	5
pH	5.8	6.0	6.2	6.4	6.6	6.8	7.0	7.2	7.4	7.6	7.8	8.0

8. Sorensens's phosphate buffer (pH 7.0)

Stock solutions

A. $KH_2 PO_4$ = 9.078g in water and make up to 1000ml.

B. $NA_2 HPO_4$ = 11.86g in water and make up to 1000ml.

Mix solution A and B in ration 4:6.

9. Potassium phosphate NaOH buffer (0.05 M)

Stock solutions

A. 0.2M $KH_2 PO_4$ = 27.22g in water and make up to 1000ml.

B. 0.2M NaOH

Mix x ml of A with y ml of B and dilute to 20ml with water.

x ml	5	5	5	5	5	5	5	5	5	5	5	5
y ml	0.36	0.56	0.81	1.16	1.64	2.24	2.91	3.47	3.91	4.24	4.45	4.61
pH	5.8	6.0	6.2	6.4	6.6	6.8	7.0	7.2	7.4	7.6	7.8	8.0

10. Maleate buffer*

Stock solutions

A. 0.2M Solution of acid sodium maleate = 8.0g of NaOH plus 23.2g of maleic acid or 19.6g of maleic anhydride in water and make up to 1000ml.

B. 0.2M NaOH

Mix 50ml of A with x ml of B and dilute to 200ml with water.

x ml	7.2	10.5	15.3	20.8	26.9	33.0	38.0	41.6	44.4
pH	5.2	5.4	5.6	5.8	6.0	6.2	6.4	6.6	6.8

* Temple, W. (1929). *J. Am. Chem.* Soc., 51, 1754.

11. Succinate buffer*

Stock solutions

A. 0.2M Solution of succinic acid = 23.6g in water and make up to 1000ml.

B. 0.2M NaOH

Mix 25ml of A with x ml of B and dilute to 200ml with water.

x ml	7.5	10.0	13.3	16.7	20.0	23.5	26.5	30.3	34.2	37.5	40.7
pH	3.8	4.0	4.2	4.4	4.6	4.8	5.0	5.2	5.4	5.6	5.8

* Gomori, G.

12. Tris (hydroxymethyl) aminomethane-HCI buffer (0.1M)

Stock solutions

A. 0.2M Solution of trist (hydroxymethyl) aminomethane = 24.23g in water and dilute to 1000ml.

B. 0.2M HCI

Mix 50ml of A with x ml of B and dilute to 100ml with water.

x ml	43	41	39	34	29	24	18	13	9.5
pH	7.2	7.4	7.6	7.8	8.0	8.2	8.4	8.6	8.8

13. Tris (hydroxymethyl) aminomethane-maleate (Tris-maleate) buffer*

Stock solutions

A. 0.2M Solution of Tris acid maleate = 24.2g of Tris (hydroxymethyl) aminomethane plus 23.2g maleic acid or 19.6g of maleic anhydride in water and make up to 1000ml

B. 0.2M NaOH

Mix 50ml of A with x ml of B and dilute to 200ml with water.

x ml	7.0	10.8	15.5	20.5	26.0	31.5	37.0	42.5	45
pH	5.2	5.4	5.6	5.8	6.0	6.2	6.4	6.6	6.8
x ml	48.0	51.0	54.0	58.0	63.5	69.0	75.0	81.0	86.5
pH	7.0	7.2	7.4	7.6	7.8	8.0	8.2	8.4	8.6

* Gomori, G. (1948). *Proc. Soc. Exptl. Biol. Med.*, 68, 35.

•••

Bibliography

- Brush, S.B. (1995) *In situ* conservation of landraces in centers of crop diversity. *Crop Sci.*, 35, p 346-354.
- CIP-UPWARD (2003) *Conservation and Sustainable Use of Agricultural Biodiversity*; A Source book. International Potato Center- Users Perspective with Agricultural Research and Development, Los Banos, Laguna, Philippines.
- *Climate Change and Plant Genetic Resources* (1990) eds. M. Jackson, B.V. Ford Lloyd and ML Parry. Belhaven Press, London.
- Cromarty, A.S., R.H. Ellis and E.H. Roberts1982. *The Design of Seed Storage Facilities for Genetic Conservation.* IBPGR, Rome.
- Crossa, J., C. M. Hernandez, P. Bretting, S. A. Eberhardt and S.Taba. (1993) Statistical genetic considerations for maintaining germplasm collections. *Theoretical and Applied Genetics*, 86: 673-678.
- Darlymple, D.(1986) *Development and spread of high-yielding wheat varieties in developing countries.* USAID, Washington, DC.
- Delouche, C. and Baskin, C.C. (1973)Accelerated ageing techniques for predicting the relative storability of seed lots. *Seed Science and Technology*, 1:427-452.

- Duvick, D.N.(1984) Genetic diversity in major farm crops on the farm and in resource. *Economic. Botany* 38 (2) pp161-178.
- Evenson, R.E. and Gollin, D. (1997) Genetic resources, international organizations and rice varietal improvement. *Economic Development and Cultural Challenge* 45(3), 471-500.
- Evenson, R.E., Gollin, D. Santaniello, V. (1998) Agricultural values of plant genetic resources In: *Agricultural Values of Plant Genetic Resources* (eds.) R.E.Evenson, , D.Gollin, V. Santaniello, 1-28p, FAO, CABI, Wallingford, UK.
- Engels, J.M.M. and R. Ramnatha Rao, eds. *Regeneration of Seed Crops and their Wild Relatives.* Proceedings of a Consultation Meeting, 4-7 December 1995, ICRISAT, Hyderabad, India. International Plant Genetic Resources Institute Rome, Italy.
- Ellis, R.H., Hong, T.D. and Roberts, E.H. 1985. Handbook of seed technology for Genebank . Vol. I. Principles and methodology. IBPGR, Rome.
- Frankel,O.H. 1984. Genetic perspectives of germplasm conservation. Pp161-170 in W. Arber, R.Illmensee, W.J.Peacock, P. Starlinger (eds.) *Genetic Manipulation: Impact on Man and Society.* Cambridge Univ. Press, Cambridge
- FAO, 1996 *Report on the state of the world's plant genetic resources for food and agriculture.*75pp, Food and Agriculture Organization of the United Nations, Rome, Italy.
- Hammer, K.1993b Generosion aus Genbank- Sicht. Vortr. Pflanzenzuchtg, 25: 140-148.
- Hammer, K., Knupffer, H., Xuvelli, L, and Perrino, P.1996a Estimating genetic erosion in land races-two case studies. Genetic Resources and Crop Evolution 43: 329-336.
- Hammer, K.1998 Agrarbioiversitat und pflanzegenetische Ressourcen-Herausforderung and Losungsansatz. Schriften Genet.Resour, 10:98
- Hawkes, J.G. (1983) *The Diversity of Crop Plants.* Harvard University Press, Cambridge, MA.
- Hanson, J. 1985 Procedures for Handling Seeds in Genebanks. Practical manuals for Genebanks: No. 1 IBPGR, Rome.
- Harlan, J.R. 1971Agricultural origin: centers and non centers. Science 174: 468-474.

- Harlan, J.R. and de Wet, J.M.J. (1971) Towards a rational classification of cultivated plants. *Taxon* 20: 509-517.
- Harlan, J.R. (1992) *Crops and Man*. American Society of Agronomy Inc., and Crop Science Society of America,Inc., Madison, Wisconsin.
- Holden, JHW, Peacock, J. and Williams, T. 1993. *Genes, Crops and the Environment*. Cambridge Univ. Press, Cambridge
- IPGRI. 1994. *Genebank Standards*. Food and Agriculture Organization of the United Nations / International Plant Genetic Resources Institute, Rome.
- International Seed Testing Association. 1985. International rules for seed testing. *Seed Sci. & Technol*. 13:338, 493-495.
- ISTA. 2005a. *International Rules for Seed Testing*. Edition 2005. International Seed Testing Association, Bassersdorf, Switzerland.
- IUCN. 1980. World Conservation Strategy. International Union for the Conservation of Natural Resources. Gland, Switzerland.43pp.
- Oldfield, M.L. (1989) The value of conserving genetic resources. Sinauer Associates, Sunderland, Massachusetts.
- Plucknett, D.L., Smith, N.J.H., Williams, J.T. Murthy Anishetty, N. (1987) Wild species: The wider genepool. In: *Gene banks and the World's Food*, pp155-170, Princeton University Press, Princeton, New Jersey.
- Prescott -Allen, C. and Prescott -Allen, R (1986) The first resource: wild species in the North American economy. Yale University, New Haven.
- Roberts and Ellis, 1984 *Crop Genetic Resources: Conservation and Evaluation* (Eds., Holden J.H.W. and J.T.George) Allen and Unwin, London, Pp18-37.
- Smale, M. 1998 Indicators of varietal diversity in bread wheat grown in developing countries. In: *Agricultural Values of Plant Genetic Resources* (eds.) Evenson, R.E., Gollin, D. Santaniello, V. 1998 FAO, CABI, Wallingford, UK.
- Schucking, H. and Anderson, P. (1991) Voices unheard and unheeded. In: Shiva V. et al.(eds) *Biodiversity: Social and Ecological Perspectives*. Zed books Ltd, London and New Jersey, with World Rainforest Movement, Penang, Malaysia.

- Wood, D. & Lenne' J.M.(1999) *Agrobiodiversity : Characterization, Utilization and Management.* CABI, UK
- United Nations Environment Program (UNEP) 1995. Global Biodiversity Assessment. Cambridge University Press, Cambridge.
- Vavilov, N.I.(1926) Studies on the origin of cultivated plants. *Bull. Appl. Bot. Genet. Pl .Breed.* 16:1-248.
- Virchow,D.1999 Efficient Conservation of Crop Genetic Diversity-Theoretical Approaches and Emperical Studies. Springer, New York
- World Conservation Monitoring Center (1992). Global Biodiversity Status of the Earth's Living Resources. Chapman and Hall, London.
- Wilson, E.O. (ed.)(1988) *Biodiversity* National Academy Press, Washington, DC.
- Zeven, A.C. 1996 Results of activities to maintain landraces and other materials of some European countries in situ before 1996 and what we can learn from them. Genetic Resources and Crop Evolution 43: 337-341.
- Zeven, A.C. and Zhukovsky 1975. Dictionary of cultivated plants and their centers of diversity. PUDOC, Wageningen, 219p.

Literature Cited

- Arora RK and Nayar ER. 1984 Wild Relatives of Crop Plants in India NBPGR Sci Mongr. No 7 National Bureau of Plant Genetic Resources pp 90.
- Harlan J.R. 1975 Crops and Man Amar Soc of Agronomy, Crop Sci Soc of America, Madison, Wisconsin USA. pp 195.
- Harlan J.R. & J M J de Wet 1971 Towards a rational classification of cultivated plants Taxon .20 :50 9-5A.
- Hoyt. E. 1988 Conserving the Wild relatives of crops NBPGR FUCON WWF pp 1-45.

•••